Lecture Notes in

Topics in Path Integrals and String Representations

Lecture Notes in

Topics in Path Integrals and String Representations

Luiz C L Botelho

Federal Fluminense University, Brazil

NEW JERSEY • LONDON • SINGAPORE • BEIJING • SHANGHAI • HONG KONG • TAIPEI • CHENNAI

Published by

World Scientific Publishing Co. Pte. Ltd.
5 Toh Tuck Link, Singapore 596224
USA office: 27 Warren Street, Suite 401-402, Hackensack, NJ 07601
UK office: 57 Shelton Street, Covent Garden, London WC2H 9HE

Library of Congress Cataloging-in-Publication Data
Names: Botelho, Luiz C. L.
Title: Lecture notes in topics in path integrals and string representations /
Luiz C.L. Botelho (Federal Fluminense University, Brazil).
Description: New Jersey : World Scientific, 2017.
Identifiers: LCCN 2016054695 | ISBN 9789813143463
Subjects: LCSH: Path integrals. | Integral representations. | Integration, Functional. |
Gauge fields (Physics)
Classification: LCC QC174.17.P27 B6745 2017 | DDC 530.12--dc23
LC record available at https://lccn.loc.gov/2016054695

British Library Cataloguing-in-Publication Data
A catalogue record for this book is available from the British Library.

Printed in Singapore

To my aphrodites, Mermaids

and ninphes, including "Hanna Nelma" – my wife

Preface

Since the years 1978–1984, the search for an analytical formalism for handling Non Abelian Gauge Theories through String Path integrals has been pursued with some success by strong interactions Theoretical-Mathematical Physicists. The basic idea is very simple: write loop wave equations for the $SU(\infty)$ quantum Wilson loop, at least in a formal Mathematical level in relation to the rigorous mathematical meaning of the loop functional derivatives and to the rigorous (perturbative or not) Euclidean quantum Yang-Mill measure – and try to solve it – at least for some classe of suitable surfaces; through a string path integral.

These ideas have been throughly exposed in my previous monograph entitled Methods of Bosonic and Fermionic Path Integral Representations – Continuous Random Geometry in Quantum Field Theory. Nova Science Publishers - Inc. NY, 2009 and references therein.

My aim in this another set of informal lecture notes on the subject is to present author's original research material and developments on this problem done after 2007; Mainly based on revised and amplified on author's published papers on international scholar journals during 2007–2015.

The methodology used to write my set of lecture notes is the same of our previous lecture notes sets/research monographs ([1]): My exposition is intend to be quick and specially expected to be thought – provoking. It is informal, with graduate students on mind and the application – oriented style is expected also to estimulate the involviment of Mathematicians in solving and analyzing the problem on the subject.

Cumbersome use of English and certainly types and spelling mistakes reflects the author's limitations on the using of advanced English grammar. The reader's criticism and comments are welcome.

Luiz Carlos Lobato Botelho

CNPq- Visiting Research – Grant 164438/2015-7 – IMECC-Universidade de Campinas, São Paulo, Brazil.

[1] Botelho, Luiz C.L., Methods of Bosonic and Fermionic Path Integrals Representations, Continuum Random Geometry in Quantum Field Theory, Nova Science Publishers, Inc., New York-USA, ISBN: 978-1-60456-068, (2009).

PS. Special thanks to Professor Waldyr Rodriguez by Sponsoring my research stay at UNICAMP.

Contents

Chapter 1

Bosonic Free Strings and Non-supersymetric $QCD(SU(\infty))$: A constant gauge field path integral study

We study in a reduced dynamical model for $QCD(SU(\infty))$, defined by constant gauge fields Yang-Mills path integral, several concepts on the validity of string representations on $QCD(SU(\infty))$ and the confinement problem.

1.1 Introduction

In the last years approaches have been pursued to reformulate non supersymetric quantum chromodynamics as a String Theory ([1], [2], [3]) and thus handle the compound hadron structure in the QCD model for strong interactions ([3]). The common idea of all those attempts is to represent the full quantum ordered non supersymetric phase factor as a string path integral, which certainly takes into account more explicitly the geometrical setting of the non abelian gauge theory than its usual description by gauge potential.

Other main protocol to achieve such string representation for the wilson loop operator in QCD is to use the still not completely understand large number of colors of t'Hooft for non supersymetric quantum Yang-Mills theory.

It is the purpose of this chapter to evaluate the static potential between two static charges with opposite signal on the approach of an effective reduced quantum dynamics of Yang-Mills constant-gauge fields ([3]) these results surely expected to be relevant for the validity of the old conjecture of E. Witten about a $QCD(SU(\infty))$ dynamics dominated by a constant gauge $SU(\infty)$ master field configuration ([4], [5], [6]). These studies are presented in section 2 of this chapter. In section 3, we present the relevant

$QCD(SU(\infty))$ loop wave equation for our reduced model of constant – gauge fields for $QCD(SU(\infty))$ and suggest that, a free bosonic string as solution for this reduced Loop Wave Equation ([7]). We continue with our study and present also a detailed calculation of the quark-antiquark static potential from a one-loop approximation on the Regge slope string constant directly from the well-known Nambu-Goto string path integral ([8]). [1]

Finally in section 4, we present also somes studies on the dynamical aspects of this framework of constant field model by presenting path integral studies on evaluation of vectorial-scalar color singlet quark currents ([6],[8]).

Before to proceed, let us firstly reproduce two enlightneen discourses on the present day problem of handle quantitatively Yang-Millls fields outside the lattice approximation

1 – Quoted from A. Jaffe and Eduard Witten.

Classical properties of non abelian gauge theory are within the reach of established mathematical methods, and indeed, classical non abelian gauge theory has played a very important role in pure mathematics in the last twenty years, especially in the study of three- and four-dimensional (C^{∞}-differentiable) manifolds.

On the other hand, one does not yet have a mathematically complete example of a quantum gauge theory in four-dimensional space-time, not even a non abelian quantum gauge theory in four-dimensional.

2 – Related to the pure string holographic approach based on the Maldacena conjecture and Super String Theory it appears interesting to cite V. Rivasseu (Math-ph/0006017) about the general philosophy underlying supersymmetric strings.

Today the main strean of theoretical physics holds the view that field theory is only an effective (approximated!) theory and that superstring or its variant, M-theory are the best candidate for a fundamental global theory of nature (including QCD).

However this superstring theory has not yet received direct experimental confirmation; it has (surely) opened up a new interface with mathematicians, mostly centered around concepts and ideas of geometry and topology (of C^{∞}-manifolds), with algebra and geometry dominating over analysis and calculational aspects.

Fortunately there is Lattice Gauge Theory, which although has remained largely phenomenological, it has produced somewhat "precise" results on Experimentall Hadron mass spectroscopy as it has been pointed

[1]It is worth to call attention that our constant gauge fields at $SU(\infty)$ are not the rigorous continuum version of the one-plaquette the Eguchi-Kawai lattice model.

out by F. Wilczek (Nature 456,449, 2008). Being enough for that by just taking some mesons (π, μ, Σ) mass inputs, even if in the context of QED one needs as input only the fine structure $\alpha = \frac{e^2}{4\pi\hbar c} \sim \frac{1}{137}$.

As again in this chapter we try to implement the QED one universal protocol for QED, by taking now our reduced model as the QCD effective theory at large N_c and as universal imput parameter, the Gluonic condensate $\langle 0|\mathrm{Tr}(F^2)|0\rangle_{SU(\infty)}$ ([7]).

1.2 The Static Confining Potential for the Eguchi Kawai Model on the Continuum

The basic gauge-invariant observable on probing the non-perturbative vacuum ([1]) of $SU(N)$ Euclidean Yang-Mills bosonic field theory on R^4 is the Wilson loop quantum average

$$\langle W[C]\rangle = \frac{\int d\mu[A]\, W[C]}{\int d\mu[A]} \tag{1.1}$$

where the loop parallel transport $SU(N)$-valued matrix is given by

$$W[C] = \frac{1}{N}\mathrm{Tr}_{SU(N)}\mathbb{P}\left\{\exp\left[ig\oint_C A_\mu(x)\, dx_\mu\right]\right\} \tag{1.2}$$

and $d\mu[A]$ denotes the Yang-Mills path-integral measure given formally by the Feynman prescription

$$d\mu[A] = \left(\prod_{x\in R^4}(dA(x))\right) \times \exp\left(-\frac{1}{4}\int_{R^D}\mathrm{tr}(F^2_{\mu\nu}(A))(x)d^4x\right). \tag{1.3}$$

The Gauge connection $A_\mu(x)$ and the field strenght $F_{\mu\nu}(x)$ are explicitly given by

$$\begin{aligned} F_{\mu\nu}(x) &= (\partial_\mu A_\nu - \partial_\nu A_\mu + ig[A_\mu, A_\nu]_-)(x) \\ A_\mu(x) &= A^a_\mu(x)\lambda_a \end{aligned} \tag{1.4}$$

The $SU(N)$ generators $\{\lambda_a\}_{a=1,\ldots,N^2-1}$, are supposed to be Hemiteans and satisfying the will known matrix relationship below

$$\begin{aligned} \left[\lambda^a, \lambda^b\right] &= i f_{abc}\, \lambda_c \\ \mathrm{Tr}\left(\lambda_a\, \lambda_b\right) &= \frac{\delta_{ab}}{2} \\ f^{abc}\, f^{dbc} &= N\, \delta^{ad} \end{aligned} \tag{1.5}$$

In our proposal for the Euchi-Kawai model on continuum, we introduce the space-time trajector $C_{(R,T)}$ of a static quark-antiquark pair, separated apart a distance R with a (Euclidean) temporal evolution $0 \le x_0 \le T$ ($R^4 = \{(x_0, \vec{x}), \vec{x} \in R^3)$.

We face thus the problem of evaluating the path integral eq. (1.1) for constant gauge fields configurations at the large N_c limit ([5], [6]) and at the physical space-time R^4.

Let us briefly review our previous framework.

We firstly consider the full infinite volume space-time R^4 reduced to a finite volume space-time Ω. This step has the effect to turns our "reduced" path integrals mathematically well defined.

This finite volume space-time is supposed to be formed by the supperposition of p four-dimensional hypercubes of caracteristic volume ("size") $V = L^4$.

The area $S[C_{(R,T)}]$ enclosed by our rectangle $C_{(R,T)}$ is such that $S[C_{R,T)}] \sim q\,L^2$ for large p. Obviously $\big(S[C]\big)^2 \le \mathrm{vol}(\Omega)$. So ours integers q and p should satisfy $q \le \sqrt{p}$, an important bound to be kept on mind on what follows.

Our improved large N-limit will be defined in such way by already taking into account the basic phenomena of QCD-Yang-Mills dimensional transmutation of the strong coupling constant, a fully non perturbative phenomena similar to the Higgs mass mechanism on Weinberg-Salan theory. We, thus, define the effective $SU(\infty)$ coupling constant through the relationship for our space-time of finite volume

$$\lim_{N\to\infty} \left(\frac{g^2 N}{L^2}\right)\left(\frac{q}{p}\right) < \infty. \tag{1.6}$$

It is worth observe that for $p \to \infty$ the infinite volume limit should be taken according the underlying $N_c \to \infty$ limit. Namely

$$\frac{q}{p} \equiv a \to 0 \tag{1.7}$$

$$\lim_{\substack{N\to\infty \\ a\to 0}} g^2 \left(\frac{Na}{L^2}\right) = (g_\infty)^2_{\text{dim}} \tag{1.8}$$

Here L^2 is a physical finite area parameter expected to be related to the domain area size of the famous QCD spaghetti vacuum ([7]).

After these preliminaries remarks, we must solve the invariant constant gauge field $SU(\infty)$ matrix path integral below written:

$$W_{SU(\infty)}\left[C_{(R,T)}\right] = \Bigg\{ \lim_{N\to\infty} \frac{1}{W(0)} \times \left[\int_{-\infty}^{+\infty} \left(\prod_{a=1}^{N^2-N} \prod_{\mu=0}^{D-1} d\,A_\mu^a \right) \right]$$
$$\times\, \Delta_{Fp}[A_\mu] \exp\left(+ \frac{g^2}{4} V \,\mathrm{Tr}\left(\left[A_\mu, A_\nu\right]_-^2 \right) \right)$$
$$\times \left(\exp \frac{|g^2 S[C_{(R,T)}]|^2}{2} \cdot \frac{1}{N} \mathrm{Tr}_{SU(N)} \left([A_0, A_1]^2 \right) \right] \Bigg) \Bigg\}. \tag{1.9}$$

In order to evaluate the $SU(N)$-invariant constant Gauge field path integral eq. (1.9), we use the Bollini-Giambiagi Cartan matrix decomposition ([5])

$$A_\mu = B_\mu^a\, H_a + G_\mu^b\, E_b \tag{1.10}$$

where the Cartan basis $\{H_a, E_a\}$ of the $SU(N)$ Lie algebra possesses the special calculations properties ([5], [6])

a) For $a, b = 1, 2, \ldots, N-1$

$$\left[H_a, H_b\right]_- = 0. \tag{1.11}$$

b) For $b = \pm 1, \ldots, \pm \dfrac{N(N-1)}{2}$

$$\left[H_a, E_b\right]_- = r_a(b)\, E_b. \tag{1.12}$$

c) For $a = 1, 2, \ldots, \dfrac{N(N-1)}{2}$.

$$\left[E_a, E_{-a}\right] = \sum_{\ell=1}^{N-1} r_c(a)\, H_a. \tag{1.13}$$

d) For $a \neq -b$, $a, b = \pm 1, \ldots, \pm \dfrac{N(N-1)}{2}$

$$\left[E_a, E_b\right]_- = N_{ab}\, E_{a+b}. \tag{1.14}$$

In this distinguished Lie Algebra basis, one can easily fix the Gauge on the $SU(N)$-valued invariant matrix path integral by simply choosing all the N-abelian components B_μ^a on the connection eq. (2.10) to be vanished. Namely

$$B_\mu^a = 0. \tag{1.15}$$

It is expected thus that the Faddev-Popov term $\Delta_{Fp}[A_\mu]$ should be quenched at the $N \to \infty$ limit eq. (1.6)-eq. (1.8) i.e. ([5])

$$\lim_{N\to\infty} \Delta_{Fp}[A_\mu] \to 1 \tag{1.16}$$

Any way we take the Faddev-Popov quenched determinant to unity in this sort of approximate evaluation of ours. $SU(\infty)$-matrix valued invariant path integral (note that procedure to evaluate degrees of Freedom reduced path integrals is usually implemented when one handles for instance, fermions degrees of Freedon on $SU(N)$ lattice path integral ([3]). We will adopt such procedure here).

By assembling all the above results one gets the following outcome eq. (1.9) defined now by $SU(N)$ constant gauge field configurations for a general euclidean space-time R^D from now on

$$W\left[C_{(R,T)}\right] = \frac{1}{W[0]} \times \left\{ \int_{-\infty}^{+\infty} \left[\prod_{a=1}^{N^2-N} \prod_{\mu=0}^{D-1} dG_\mu^a \right] \right\}$$

$$\exp\left\{ +\frac{1}{2} G_\mu^a\, G_\nu^b\, G_\mu^c\, G_\nu^d\, \mathcal{L}_{abcd} \left[g^2 V + \left(\delta_{\mu 0}\, \delta_{\nu 1} \frac{[g^2 S][C_{(R,T)}]}{(\frac{N}{2})} \right) \right] \right\} \tag{1.17}$$

[2]

The above matrix valued path integral can be easily exactly evaluated through re-scalings, at large N: Namelly (see Appendix A for details).
a) For $\mu \neq 0, 1$: $G_\mu^a \to G_\mu^a [g^2 V]^{-\frac{1}{4}}$.
b) For $\mu = 0, 1$;

$$G_\mu^a \to G_\mu^a \left[g^2 + \frac{[g^2 S[C_{(R,T)}])^2]}{(\frac{N}{2})} \right]^{-1/4} \tag{1.18}$$

And leading thus to the exactly result

$$W\left[C_{(R,T)}\right] = \lim_{N\to\infty} \left\{ \left[\left(g^2 V + \frac{(g^2 S[C_{(R,T)}])^2}{\frac{N}{2}} \right)^{-\left(\frac{N^2-N}{2}\right)} \right.\right.$$

---

[2]

$$\mathcal{L}_{abcd} = \left(\sum_{i,\ell=1}^{N-1} r_i(a)\, r_\ell(c)\, \delta_{i\ell}\, \delta_{c_i - d}\, \delta_{c_i - b} \right)$$
$$+ \left(N_{ab}\, N_{cd}\, (1 - \delta_{a_i - b})(1 - \delta_{c_i - d}) \delta_{a+b,-(c+d)} \right)$$

$$\times \quad (g^2V)^{-\frac{[N^2-N)(D-2)}{4}}\Big]\Big/(g^2V)^{-\frac{(N_2-N)D}{4}}\Big\} \tag{1.19}$$

It yields thus, the following $N \to \infty$ limit

$$W[C_{R,T}] = \lim_{N\to\infty}\left(1+\frac{g^2S^2[C_{R,T)}]}{\left(\frac{N}{2}\right)V}\right)^{\frac{-N(N-1)}{2}}$$

$$= \quad \lim_{N\to\infty}\left\{\exp\left[-\frac{g^2(N-1)L^2}{L^2}\left(\frac{S^2}{V}\right)\right]\right\} \tag{1.20}$$

For $D = 4$, we have on the context of our proposed $SU(\infty)$ infinite volume limit eqs. (1.6)-(1.8), our R^4 Wilson Loop "string" behavior.

$$W[C_{(R,T)}] = \exp\left\{-\frac{g^2(N-1)}{L^2}L^2\left(\frac{q^2L^4}{pL^4}\right)\right\}$$

$$\overset{(N\to\infty)}{\sim} \quad \exp\left\{-\frac{g^2N}{L^2}\left(\frac{q}{p}\right)(qL^2)\right\}$$

$$\overset{(N\to\infty)}{\sim} \quad \exp\left\{-(g_\infty)^2\,RT\right\} \tag{1.21}$$

From eq. (1.21), one obtains the confining quark-antiquark potential

$$V(R) = \lim_{T\to\infty}\left(-\frac{1}{T}\,\ell g(W[C_{R,T)}])\right) = (g_\infty)^2\,R \tag{1.22}$$

leading to an attractive constant force "biding" the static pair of quarks as originally obtained by K. Wilson on his lattice gauge - modelling ([3])

1.3 The Luscher correction to inter quark potential on the reduced model

In this section we intend to show that our proposed $SU(\infty)$ constant gauge field theory leads to a free string theory path-integral. We thus evaluate explicitly through the string path-integral the next non-confining corrections to the quark-antiquark potential eq. (2-22).

In order to argument an effective low energy QCD string representation in this model, we are going to consider the loop have equation ([1]) for constant gauge fields already on the continuum at large N limit.

Let us thus firstly consider general loops $C_{x\,x} = \left\{\left(X_\mu(\sigma)\right)_{\mu=0,1,2,3};\, 0 \leq \sigma \leq 2\pi\right\}$ on R^4. It is well-known that formally we have the functional loop derivative

a)

$$\psi[C_{x,x}, A_\mu(x)] = \frac{1}{N} \text{Tr } \mathbb{P} \left\{ e^{ig} \oint_{C_{xx}} A_\mu(X(\sigma)) dX_\mu(\sigma) \right\} \tag{1.23}$$

b)

$$\frac{\delta}{\delta X_\mu(\overline{\sigma})} \quad \psi[C_{xx}, A_\mu(x)] = \frac{ig}{N} \text{Tr } \mathbb{P} \left\{ F_{\mu\nu}(X_x(\overline{\sigma})) \frac{dX^\nu(\overline{\sigma})}{d\overline{\sigma}} \right) \times \exp\left(ig \oint_{C_{xx}} A_\mu(X(\sigma)) dX^\mu(\sigma) \right) \right\} \tag{1.24}$$

c)[3]

$$\frac{\delta^2}{\delta X_\mu(\overline{\sigma}) \delta X^\mu(\overline{\sigma})} \quad \psi_{SU(N)} \left[C_{xx}, A_\mu(x) \right]$$

$$= \left\{ \left[\frac{1}{N} ig \text{ Tr } \mathbb{P} \left\{ \overbrace{(\nabla F_{\mu\nu})(X^\beta(\overline{\sigma})) \frac{dX^\nu}{d\overline{\sigma}}}^{=\frac{\delta}{\delta X_\mu(\overline{\sigma})}(F_{\mu\nu}(X(\overline{\sigma})))} (\overline{\sigma}) \exp\left(ig \oint_{C_{xx}} A_\mu dX_\mu \right) \right) \right] \right.$$

$$+ \left[\frac{(ig)^2}{N} \text{ Tr } \mathbb{P} \left\{ (F_{\mu\nu} F_{\mu\nu})(X^\beta(\overline{\sigma})) \right. \right.$$

$$\left. \left. \times \left(\frac{dX^\rho}{d\overline{\sigma}} \cdot \frac{dX^\rho}{d\overline{\sigma}} \right) (\overline{\sigma}) \exp\left(ig \oint_{C_{xx}} A_\mu dX_\mu \right) \right) \right] \right\} \tag{1.25}$$

For constant gauge fields configurations the first term of the right-hand side of eq. (1.25) $\left(\frac{\delta}{\delta X_\mu(\overline{\sigma})} (\text{ constant } F_{\mu\nu}) = 0! \right)$ vanishes identically. So, after taking the path integral average of eq. (3.25) through the path integral of constant gauge fields configurations eq. (1.9) and considering the usual path integral factorization of a product of gauge invariant observable at $SU(\infty)$, together with the formation of non-vanishing value of the Yang-Mills energy on the non-trivial QCD vacuum one gets finally the following loop wave equation for the quantum Wilson Loop our Loop in our reduced $SU(\infty)$ gauge theory on R^4.

$$\int_0^{2\pi} \left[\left(\frac{\delta^2}{\delta X_\mu(\overline{\sigma}) \, \delta X^\mu(\overline{\sigma})} \right) \psi_{SU(\infty)} \left[X^\mu(\sigma) \right] \right] d\overline{\sigma}$$

[3]The central hypothesis about the Constant Gauge Fields $QCD(S \cup (\infty))$ stringy vacuum:

$$\langle \Omega_{\text{Vac}} | (F_{\mu\nu} F^{\mu\alpha})(x) | \Omega_{\text{Vac}} \rangle = \int_{\nu\alpha} \langle \Omega_{\text{Vac}} | (F_{\mu\nu} F^{\mu\nu}) | \Omega_{\text{Vac}} \rangle.$$

$$(\geq 0)$$

$$= \left(-\left(g_\infty\right)^2_{(0)} \langle 0|F^2|0\rangle_{SU(\infty)}\right) \times \left(\int_0^{2\pi} \left|X'_\mu(\overline{\sigma})\right|^2 \times \psi_{SU(\infty)}\left[X^\mu(\sigma)\right]\right] d\overline{\sigma}\right) \tag{1.26}$$

Here, we have the $SU(\infty)$ Euclidean Gauge Theory parameters identification a)

$$\left(g_\infty\right)^2_{(0)} = \lim_{N\to\infty} (g^2 N)$$

b)

$$\delta^{\rho\alpha} \langle 0|F^2|0\rangle_{SU(\infty)} = \lim_{N\to\infty} \left(\frac{1}{N} \operatorname{Tr} \langle 0| \int d^D x(F_{\mu\rho}(x)\, F^{\mu\alpha}(x))\,|0\rangle\right) < 0 \tag{1.27}$$

By comparing the above parameters with those coupling constants of the static case,one has the following identification for the Spaghetti QCD non perturbative broken scale invariance vacuum effective area domain with the QCD value condensate

$$\frac{1}{a^{\text{eff}}} = (-\langle 0|F^2|0\rangle) > 0. \tag{1.28}$$

A result already expected ([7]).

At this point we point out that the reduced loop wave equation is the same of a free Bosonic string theory with the string Regge slope identification with the reduced Gauge theory at $SU(\infty)$

$$\frac{1}{(2\pi\alpha')^2} = -\left(g_\infty\right)^2_{(0)} \langle 0|F^2|0\rangle \tag{1.29}$$

As a consequence, one should expects the phenomenological path-integral representation between the large N_c and extreme low energy continuum $QCD(SU(\infty))$ (represented by constant $SU(\infty)$ gauge fields), with a free bosonic (creation process) string path integral on the light-cone gauge

$$\langle \psi[C_{xx}, A_\mu(x)]\rangle_{\substack{SU(\infty)\\ \text{low-energy}}} = G(C_{xx}, A)$$

$$= \int_0^\infty dA \left\{ \int_{X^\mu(\sigma,0)=0}^{X^\mu(\sigma,A)=C_{xx}(\sigma)} D^F[X^\mu(\sigma(\zeta)]\right\}$$

$$\times \exp\left\{-\frac{1}{2}\int_0^A dt \int_0^{2\pi} d\sigma \left[\left(\partial_\zeta X^\mu\right)^2 + \frac{1}{(2\pi\alpha')^2}\left(\partial_\sigma X^\mu\right)^2\right]\right\} \tag{1.30}$$

$$G_{\text{string}}\big(C_{xx},0\big)=G_{\text{string}}\big(C_{xx},\infty\big)=0 \tag{1.31}$$

At this point it is worth observe that our light-cone string path integral propagator

$$G_{\text{string}}\big(C_{xx},A\big)=\int_{X^{\mu}(\sigma,0)=C_{xx}(\sigma)} D^{F}[X^{\mu}(\sigma,\pi)]$$

$$\times\exp\left\{-\frac{1}{2}\int_0^A dt\int_0^{2\pi}d\sigma\left[\left(\partial_\zeta X^\mu\right)^2+\frac{1}{(2\pi\alpha')^2}\left(\partial_\sigma X^\mu\right)^2\right]\right\} \tag{1.32}$$

satisfies the area-Difusion euclidean Schrodinger loop functional equation:

$$\frac{\partial G_{\text{string}}\big(C_{xx},A\big)}{\partial A}$$
$$=\left\{\int_0^{2\pi}d\overline{\sigma}\left(\frac{\delta^2}{\delta X_\mu(\overline{\sigma})\delta X_\mu(\overline{\sigma})}-\frac{1}{(2\pi\alpha')^2}|X'_\mu(\overline{\sigma})|^2\right)G_{\text{string}}\big(C_{xx},A\big)\right\} \tag{1.33}$$

togheter with the boundary conditons:

$$G_{\text{string}}\big(C_{xx},0\big)=G_{\text{string}}\big(C_{xx},\infty\big)=0. \tag{1.34}$$

Let us now evaluate in details the quark-antiquark potential from the general Nambu-Goto string, path integral in R^D

$$\psi\big[C_{(R,T)}\big]=\int D^F\left[X^\mu(\sigma,\zeta)\right]$$
$$\times\exp\left\{-\frac{1}{2\pi\alpha'}\left[\int_0^T d\zeta\int_0^R d\sigma\,\sqrt{\det(h(X^\mu(\sigma,\zeta))}\right]\right\} \tag{1.35}$$

Here the orthogonal dynamical string vector position is considered as closed quantum fluctuations from the static quark-antiquark trajected by $C_{(R,T)}$ i.e.:

a)

$$X^\mu(\sigma,\zeta)=\zeta\hat{e}_1+\sigma\hat{e}_2+\sqrt{\pi\alpha'}\,Y^\mu(\sigma,\zeta)$$

b)

$$\begin{cases}Y^\mu(\sigma,\zeta\pm T)=Y^\mu(\sigma,\zeta)\\ Y^\mu(0,\pm T)=Y^\mu(0,0)\\ \mu=2,3,\ldots,D-2\end{cases}$$

c)

$$h_{00}\big(X^\mu(\sigma,\zeta)=\big(\partial_\zeta X^\mu\,\partial_\zeta X_\mu\big)(\sigma,\zeta)=1+\pi\alpha'\big(\partial_\zeta Y^\mu\,\partial_\zeta Y^\mu\big)(\sigma,\zeta)$$

d)

$$h_{01}\big(X^{\mu}(\sigma,\zeta\big) = \pi\alpha'\big(\partial_{\sigma}Y^{\mu}\,\partial_{\zeta}X^{\mu}\big)(\sigma,\zeta)$$

e)

$$h_{11}\big(X^{\mu}(\sigma,\zeta\big) = 1+\pi\alpha'\big(\partial_{\sigma}Y^{\mu}\,\partial_{\sigma}X^{\mu}\big)(\sigma,\zeta) \tag{1.36}$$

As a consequence, we have explicitly the following one-loop order approximation for the string path integral weight eq. (1.35):

$$\frac{1}{2\pi\alpha'}\sqrt{h\big(X^{\mu}(\sigma,\zeta)\big)} = \frac{1}{2\pi\alpha'}\left[1+\pi\alpha'\left(\frac{\partial Y^{\mu}}{\partial\zeta}\frac{\partial Y^{\mu}}{\partial\zeta}+\frac{\partial Y^{\mu}}{\partial\sigma}\frac{\partial Y^{\mu}}{\partial\sigma}\right)\right]\times 0((\alpha')^2) \tag{1.37}$$

As a result of substituting eq. (1.36) on eq. (1.37), one gets the following closed bosonic string Gaussian path-integral to evaluate

$$\begin{aligned}
&\psi\big[C_{(R,T)}\big] \\
&= \int_{Y^{\mu}(\sigma,\zeta\pm T)=Y^{\mu}(\sigma,\zeta)} D^{F}\big[Y^{\mu}(\sigma\zeta)\big] \\
&\times\exp\left\{\left(-\frac{RT}{2\pi\alpha'}\right)-\frac{1}{2}\int_0^T d\zeta\int_0^R d\sigma\big(Y^{\mu}(-\Delta)'_{(R,T)})Y_{\mu}\big)(\sigma,\zeta)\right\} \\
&= e^{-\frac{RT}{2\pi\alpha'}}\left(\det{}^{-\frac{(D-2)}{2}}(-\Delta)'_{(R,T)}\right)\Big)
\end{aligned} \tag{1.38}$$

where the Laplacian $-\Delta'_{(R,T)}$ on the rectangle $C_{(R,\pm 1)}$ has Dirichlet boundary conditions and considered projected out from the zero modes.

It has been evaluated fully on the literature ([8]):

$$\begin{aligned}
&\det{}^{-\frac{D-2}{2}}\big(-\Delta'_{(R,T)}\big) \\
&= \left[\left(\frac{R}{T}\right)^{(\frac{D-2}{2}}e^{+\frac{\pi T}{6R}(D-2)}\times\left(\prod_{n=1}^{\infty}\left(1-e^{-\frac{2\pi n}{R}(\frac{T}{R})}\right)^{-2(D-2)}\right)\right]
\end{aligned} \tag{1.39}$$

At this point, one can easily verify the string result for the quark-antiquark potential:

$$V(R) = \lim_{T\to\infty}\left(-\frac{1}{T}\,\ell n\,\psi\big[C_{(R,T)}\big]\right) = \frac{1}{2\pi\alpha'}R - \underbrace{\frac{(D-2)\pi}{6}}_{\text{Luscher correction}}\cdot\frac{1}{R} \tag{1.40}$$

1.4 Some path integral dynamical aspects of the reduced *QCD* as a path integral dynamics of euclidean strings

Let us start this section on dynamical aspects by writing firstly in details, the operational euclidean path integral expression for the non-relativistic Feynman propagator of a spinless particle in the presence of an external (euclidean) abelian gauge field and an external scalar potential.

As a first step let us write the Feynman propagator above cited in the euclidean space-time $R^4(\hbar = 1)$

$$G(x,y,t) = \left\langle x \left| \exp\left\{ -t \left[\frac{1}{2m} \left(-i\vec{\nabla} - \frac{e}{c}\vec{A} \right)^2 - i\,e\,\varphi + g\,V \right] \right\} \right| y \right\rangle \quad (1.41)$$

Then $\vec{A} = (A_1, A_2, A_3)$ denotes the time-dependent vectorial abelian field, φ the field potential and $V(x,,t)$, the external potential, also supposed time dependent.

The phase-space path integral is easily written as of as

$$\begin{aligned} G(x,y,t) = & \int_{\vec{X}(0)=y}^{\vec{X}(t)-x} D^F(\vec{X}(\sigma)) \int D^F[\vec{p}(\sigma)] \\ & \times \exp\left\{ +i \int_0^t \left(\vec{P} \cdot \frac{d\vec{A}}{d\sigma} \right)(\sigma)\, d\sigma \right\} \\ & \times \exp\left\{ - \left[\int_0^t \frac{1}{2m} \left(\vec{P}(\sigma) - \frac{e}{c}\vec{A}(\vec{X}(\sigma)) \right)^2 \right.\right. \\ & \left.\left. - i\,e\,\varphi(\vec{X}(\sigma),\sigma) + g\,V(X(\sigma),\sigma) \right] \right\} \end{aligned} \quad (1.42)$$

After the formal path-integral change of variable into eq. (1.42)

$$\vec{P}(\sigma) - \frac{e}{e}\vec{A}(\vec{X}(\sigma),\sigma) = \vec{Q}(\sigma) \quad (1.43)$$

we get the following result:

$$G(x,y,t) = \int_{\vec{X}(0)=y}^{\vec{X}(t)=x} D^F(\vec{X}(\sigma) \int D^F[\vec{Q}(\sigma)]$$
$$\exp\left\{ i \int_0^t \left(\vec{Q}(\sigma) + \frac{e}{c}\vec{A}(X(\sigma),\sigma'\right) \frac{d\vec{X}(\sigma)}{d\sigma} \right\}$$
$$\exp\left\{ -\frac{1}{2m} \int_0^t (\vec{Q}(\sigma))^2 \right\}$$
$$\exp\left\{ +ie \int_0^t \varphi(\vec{X}(\sigma),\sigma) d\sigma \right\}$$
$$\exp\left\{ - \int_0^t V(\vec{X}(\sigma),\sigma) d\sigma \right\} \tag{1.44}$$

After realizing the Gaussian $\vec{Q}(\sigma)$ functional integral

$$\int D^F[\vec{Q}(\sigma)] \exp\left[-\frac{1}{2m} \int_0^t \left(\vec{Q}(\sigma)\right)^2 d\sigma \right] \exp\left[i \int_0^t \vec{Q}(X(\sigma)) \frac{d\vec{X}(\sigma)}{d\sigma} \right]$$
$$= \exp\left\{ -\frac{m}{2} \int_0^t \left(\frac{d\vec{X}}{d\sigma} \right)^2 (\sigma)\, d\sigma \right\}, \tag{1.45}$$

we get our gauge invariant path integral expression for the euclidean Feynman propagator under study

$$G(x,y,t) = \int_{\vec{X}(0)=y}^{\vec{X}(t)=x} D^F[\vec{X}(\sigma)] \exp\left[-\frac{m}{2} \int_0^t \left(\frac{d\vec{X}}{d\sigma} \right)^2 (\sigma)\, d\sigma \right]$$
$$\exp\left[-g \int_0^t V(X(\sigma),\sigma)\, d\sigma \right]$$
$$\exp\left[ie \left(\int_0^t \vec{A}(\vec{X}(\sigma) \cdot \frac{d\vec{X}}{d\sigma}\, d\sigma \right) \right.$$
$$\left. + ie \left(\int_0^t \varphi(\vec{X}(\sigma),\sigma)\, d\sigma \right) \right] \tag{1.46}$$

The Gauge invariance of eq. (1.46) is under all reduced periodic Gauge transformation ($x \le z \le y$; $0 \le z' \le t$)

$$\begin{cases} \vec{A}(z,t') & = (\vec{A} + \vec{\nabla}\Lambda)(z,t') \\ \varphi(z,t') & = (\varphi + \frac{\partial\Lambda}{\partial t}(z,t') \\ \Lambda(x,t) & = \Lambda(y,0) + \frac{2\pi n}{e} \end{cases} \tag{1.47}$$

Namely

$$\exp\left\{ie\int_0^t (\overline{\nabla}\Lambda)(\overline{X}(\sigma),\sigma)\,\frac{d\overline{X}}{d\sigma}\,d\sigma + ie\int_0^t \frac{\partial\Lambda}{d\sigma}\,d\sigma\right\}$$

$$=\exp\left\{ie\int_0^t \frac{d}{d\sigma}(\Lambda(\overline{X}(\sigma),\sigma))\,d\sigma\right\}$$

$$=\exp\left\{ie\big(\Lambda(x,t)-\Lambda(y,0)\big\}\,. \tag{1.48}$$

In the euclidean quantum field case in R^D one must consider the generating fermionic case

$$Z[A_\mu]=\frac{\det{}^{\frac{1}{2}}(\ \not{D}^*(A)\ \not{D}(A))}{\det{}^{\frac{1}{2}}(\not{\partial}^*\ \not{\partial})}, \tag{1.49}$$

which can be re-write through well-known propertion loop space techniques as a loop space D-dimensionall non-relativistic propagator

$$\ell g(Z[A]/Z(A=0)) = -\frac{1}{2}\int_0^\infty \frac{dt}{t}\left\{\int_{R^D} d^D x_\mu \int_{X_\mu(0)=x_\mu}^{X_\mu(\zeta)=x_\mu} D^F[X_\mu(\sigma)]\right.$$

$$\exp\left\{-\frac{1}{2}\int_0^t\left(\frac{dX^\mu}{d\sigma}\right)^2(\sigma)\,d\sigma\right\}$$

$$\mathbb{P}_{\text{spin}}\left\{\mathbb{P}_{SU(N)}\left[\exp\left(ie\int_0^t\Big(A_\mu(X^\mu(\sigma))\Big)+\frac{ie}{4}\,[\gamma^\mu,\gamma^\nu]\,F_{\mu\nu}(X^\rho(\sigma))\,d\sigma\right)\right]\right\}, \tag{1.50}$$

where the symbols $\mathbb{P}_{\text{spin}}$ and $\mathbb{P}_{SU(N)}$ means σ-ordered matrixes indexes of the spin-color gauge connection phase factor (the fermionic Wilson loop). At very low energy region, one could consider as an effective theory, all degree of Dirac spin of the particle, non-dynamical (i.e. frozen to scalar values), or equivalently one can disregard the spin orbit shenght field coupling on eq. (1.50) $\left(\frac{ig}{\hbar}\,[\gamma^\mu,\gamma^\nu]F_{\mu\nu}(X^\beta(\sigma))\cong 0\right)$.

Let us now apply the above well-known remarks to evaluate approximately "scalar" composite operators quark-antiquark Green functions.

The effective connected generating functional for vectorial quark currents at very low energy (the strong coupling region of the underlying Massless Yang-Mills theory) is given by the following loop expression

$$\ell g\left(Z_{QCD}^{eff}\,[J_\mu]\right) = \ell g\left\langle\ \det{}^{\frac{1}{2}}(\ \not{D}^*(igA_\mu+J_\mu)\ \not{D}(igA_\mu+J_\mu))\right\rangle_{A_\mu} \tag{1.51}$$

there $\langle\ \rangle_{A_\mu}$ denotes the complete Yang-Mills path integral, A_μ the Yang-Mills field and $J_\mu(x)$ the external source of the vectorial quark currents $\big(J_\mu(x)(\overline{\psi}\gamma^\mu\psi)(x)\big)$.

On the basis of the above discusions one has the following expression for eq. (4.50), with the QCD scale Λ_{QCD} already bult in a large $SU(\infty)$ limit:

$$\ell g\left(Z^{eff}_{QCD}[J_\mu]_{\Lambda_{QCD}}\right) = -\frac{1}{2}\bigg\{\int_{1/\Lambda_{QCD}}^{\Lambda_{QCD}} \frac{dt}{t}$$

$$\times\left[\int d^D z_\mu \int_{X^\alpha(0)=X^\alpha(t)=z^\alpha} D^F(X^\alpha(\sigma)) \exp\left(-\frac{1}{2}\int_0^t \left(\frac{dX}{d\sigma}\right)^2(\sigma)\, d\sigma\right)\right.$$

$$\times\left\langle \mathbb{P}_{SU(N)}\left\{\exp ig\int_0^t \left(A_\mu \frac{dX^\mu}{d\sigma}\right)(\sigma) d\sigma\right\}\right\rangle_{A_\mu} \exp\left(i\int_0^t \left(J_\mu \frac{dX^\mu}{d\sigma}\right) d\sigma\right) \tag{1.52}$$

The vectorial N-point bilinear quark current is given by in momentum space

$$\langle(\overline{\psi}\gamma^{\mu_1}\psi)(x_1)\dots(\overline{\psi}\gamma^{\mu_N}\psi)(x_\mu)\rangle_{A_\mu}$$

$$= \frac{\delta^2}{\delta J_{\mu_1}(x_1),\dots,\delta J_{\mu_N}(x_1)}\left[\ell g\left(Z^{eff}_{QCD}(J_\mu)\right)\right]\Big|_{J_\mu\equiv 0} = G_{\mu_1\dots\mu_N}(x_1,\dots,x_N) \tag{1.53}$$

Or equivalently, after suitable Fourier momenta transforms.

$$\widetilde{G}_{\Lambda_{QCD}}(P_{\mu_1},\dots,P_{\mu_N}) = -\frac{1}{2}(i)^N\bigg\{\int_{1/\Lambda_{QCD}}^{\Lambda_{QCD}} \frac{dt}{t}\int_0^t d\sigma_1\dots\int_0^t d\sigma_N \int d^D z_\mu$$

$$\times\int_{X_\mu(0)=z_\mu}^{X_\mu(t)=z_\mu} D^F[X(\sigma)]$$

$$\times\left[\exp\left(-\frac{1}{2}\int_0^t\left(\frac{dX}{d\sigma}\right)^2(\sigma)d\sigma\right)\right.$$

$$\left.\times\left(\frac{dX_{\mu_1}}{d\sigma}(\sigma_1)\dots\frac{dX_{\mu_N}}{d\sigma}(\sigma_N)\right)\right]$$

$$\times\left(\exp\left(i\sum_{h=1}^{N} p_k^\mu X_\mu(\sigma_k)\right)\right)$$

$$\times\left\langle e^{ig\int_0^t A_\mu\, dX^\mu}\right\rangle_{SU(\infty)} \tag{1.54}$$

On the basis of eq. (1.54), one could envisage to try to evaluate eq. (1.53) through an Gaussian (euclidean) string path integral. Let us take for granted such string representation as a workable sound hypothesis on basis of our previous studies.

The key point is to evaluate in terms of the loop variable $X^\mu(\sigma)$, the following anihillation string path integral:[4]

$$W[X_\mu(\sigma), 0 \le \sigma \le t] = \int_0^\infty dA \Bigg\{ \int_{Y^\mu(\sigma,0)=X^\mu(\sigma)}^{Y^\mu(\sigma,A)=0} D^F(Y^\mu(\sigma,s))$$

$$\exp\left\{ -\frac{1}{2} \int_0^A ds \int_0^t d\sigma \left[\left(\partial_s Y^\mu\right)^2 + \frac{1}{(\pi\alpha')} \left(\partial_\sigma Y^\mu\right)^2 \right] (\sigma, s) \right\} \Bigg\} \quad (1.55)$$

In order to evaluate eq. (1.55) exactly, let us firstly consider the stndard re-scale

$$\begin{aligned} &\sigma \longrightarrow \sigma\left(\pi\alpha'\right)^{1/2} = \overline{\sigma} \\ &s \longrightarrow s \\ &Y^\mu(\sigma, s) \longrightarrow \overline{Y}^\mu(\overline{\sigma}, s) \equiv \left(\pi\alpha'\right)^{1/4}(Y^\mu(\sigma, s)) \end{aligned} \quad (1.56)$$

which formally turns the string velocity into a overall factor into the path integral weight

$$W[\overline{X}_\mu(\overline{\sigma})] = \int_0^\infty dA \Bigg\{ \int_{\overline{Y}^\mu(\overline{\sigma},0)=\overline{X}_\mu(\overline{\sigma})}^{\overline{Y}^\mu(\overline{\sigma},A)=0} D^F\left[\overline{Y}^\mu(\overline{\sigma}, s)\right]$$

$$\times \exp\left\{ -\frac{1}{(2\pi\alpha')} \left[\int_0^A ds \int_0^{t(\pi\alpha')^{1/2}} d\overline{\sigma} \left(\left(\frac{\partial \overline{Y}^\mu}{ds} \right)^2 + \left(\frac{\partial \overline{Y}^\mu}{d\overline{\sigma}} \right)^2 \right) \right] \right\} \quad (1.57)$$

After considering the "Brownian Bridge like" background loop-surface decomposition which has the meaning of considering a toroidal like fluctuating closed string world sheet $Z_\mu(\overline{\sigma}, \zeta)$ bounded by the closed quark-antiquark trajectory $\overline{X}_\mu(\overline{\sigma})$ $(\overline{X}_\mu(\overline{\sigma}+t) = \overline{X}_\mu(\overline{\sigma})$, t, fixed loop proper-time)

$$\begin{aligned} \overline{Y}_\mu(\overline{\sigma}, s) &= \overline{X}_\mu(\overline{\sigma}) \left(\frac{A-s}{A} \right) + \sqrt{\pi\alpha'}\, Z^\mu(\overline{\sigma}, s) \\ Z^\mu(\overline{\sigma}, A) &= Z^\mu(\overline{\sigma}, 0) = 0 \\ Z^\mu(\overline{\sigma} + t, s) &= Z^\mu(\overline{\sigma}, s), \end{aligned} \quad (1.58)$$

[4]If the action was $\int_0^A ds \int_0^t d\sigma[(\partial_s Y^\mu)^2 + \frac{1}{(\pi\alpha')^2}(\partial_\sigma Y^\mu)^2](s, \sigma)$, then eq. (1.56) would takes the form $\sigma \to \overline{\sigma} = \sigma(\pi\alpha')$ and $\overline{Y}^\mu(\overline{\sigma}, \overline{s}) = Y^\mu(\sigma, s)$.

one gets the regularized proper-time string propagator

$$W\left[\overline{X}_\mu(\overline{\sigma})\right] = \int_\varepsilon^\infty dA \exp\left\{ -\left(\frac{(A/3)}{2\pi\alpha'}\right)\left(\int_0^{(\pi\alpha')^{\frac{1}{2}}\cdot t}\left(\frac{d\overline{X}_\mu(\overline{\sigma})}{d\overline{\sigma}}\right)^2 d\overline{\sigma}\right)\right.$$

$$\left. -\left(\frac{1}{2\pi\alpha' A}\right)\left(\int_0^{(\pi\alpha')^{\frac{1}{2}}\, t}(\overline{X}_\mu(\overline{\sigma}))^2\, d\overline{\sigma}\right)\right\} \times \left(\det{}^{-\frac{D}{2}}\left(-\Delta_{((\pi\alpha')^{\frac{1}{2}}\, t, A)}\right)\right) \tag{1.59}$$

Just for completeness, we note the following exactly expressions for the fluctucting worl-sheet Z^μ Laplacean determinant and its Green function on the rectangle $\overbrace{\left[0, (\pi\alpha')^{\frac{1}{2}}\, t\right]}^{\sigma} \times \overbrace{[0, A]}^{\xi}$:

$$\det{}^{-\frac{D}{2}}\left(-\Delta_{((\pi\alpha)^{\frac{1}{2}} t, A)}\right) = \left(\prod_{n,m}\left[\left(\frac{2\pi n}{((\pi\alpha')^{\frac{1}{2}} t}\right)^2 + \left(\frac{2\pi m}{A}\right)^2\right]\right)^{-D/2}$$

$$= \left(\frac{(\pi\alpha')^{\frac{1}{2}} t}{A}\right)^{D/2} \exp\left(\frac{\pi}{6}\left(\frac{A}{(\pi\alpha')^{\frac{1}{2}}\, t}\right) D\right)$$

$$\times \left(\prod_{n=1}^\infty\left[1 - \exp\left(\frac{2\pi n A}{(\pi\alpha')^{\frac{1}{2}}\, t}\right)\right]\right)^{-2D} ; \tag{1.60-a}$$

$$(-\Delta)^{-1}_{((\pi\alpha')^{\frac{1}{2}}\, t, A)} \quad (\overline{\sigma}, \overline{\sigma}', s, s')$$

$$= -\frac{1}{2}\left\{\sum_{\substack{n,m\\ -\infty}}^{+\infty}\left(\frac{e^{\frac{2\pi i n(\overline{\sigma}-\overline{\sigma}')}{(\pi\alpha')^{\frac{1}{2}}\, t}}}{\left(\frac{2\pi n}{(\pi\alpha')^{\frac{1}{2}}\, t}\right)^2 + \left(\frac{2\pi m}{A}\right)^2}\right)\right.$$

$$\left.\times\left[\cos\left(\frac{2\pi m}{A}(s - s')\right) - \cos\left(\frac{2\pi m}{A}(s + s')\right)\right]\right\} \tag{1.60-b}$$

As a consequence we get for N-point euclidean scalar meson Green function after disregarding the contribution of the functional determinant eq. (1.60-a) and by considering $\pi\alpha' = 1$ from now on

$$\widetilde{G}_{(t)}(P_1^\mu, \dots, P_N^\mu)$$

$$= -\frac{1}{2} \times \left\{\int_{1/\Lambda_{QCD}}^{\Lambda_{QCD}} \frac{dt}{t}\int_0^t d\sigma_1 \dots \int_0^t d\sigma_N\left[\int_\varepsilon^\infty dA\right.\right.$$

$$\left.\left.\times F\big((P_k^\mu \cdot P_{k'}^\mu), A, T, \{\sigma_1, \dots, \sigma_N\}\big)\right]\right\}. \tag{1.61}$$

Where the quark-antiquark harmonic oscillator form factor coming from eqs. (1.59), eq. (1.54) (with for notation simplicity $\pi\alpha' = 1$ and by considering the scalar case $\frac{dX^{\mu_1}(\sigma_1)}{d\sigma} \dots \frac{dX^{\mu_N}(\sigma_N)}{d\sigma} \to 1$) is given explicitly by the result:

$$
\begin{aligned}
& F\big((P_k^\mu \dot{P}_{k'}^\mu), A, t, \{\sigma_1, \dots, \sigma_\mu\}\big) \\
&= \left(\frac{(3+2A)\sqrt{\frac{2}{A}}}{6\pi \sin h\big(\sqrt{\frac{2}{A}}\, t\big)} \right)^{D/2} \times \exp\left\{ - \frac{3\sqrt{A}}{\sqrt{2}(3+2A)\sin h\big(\sqrt{\frac{2}{A}}\, t\big)} \right. \\
&\left. \times \left[\sum_{\substack{k=1 \\ k'=1}}^{N} (P_k^\mu \cdot P_{k'}^\mu) \left(\sin h \left(\sqrt{\frac{2}{A}} \right) (t - \sigma_k) \right) \times \sin h \left(\sqrt{\frac{2}{A}} \right) \sigma_{k'} \right] \right\}
\end{aligned} \tag{1.62}
$$

It is very important to remark that our "toy model" given by eq. (1.62) has the correct structure to generate a Lorentz-invariant scattering amplitude, after continuation to Minkowski space, on the light of the Hall and Wightman theorem ([9]) ($2\pi\alpha' = 1$). Namely:

$$
G_{(\Lambda_{(QCD,t)}}\left(P_1^\mu, \dots, P_N^\mu\right) = F_{(\Lambda_{QCD})}\left(P_\mu^i \cdot P_\mu^k\right) \tag{1.63}
$$

One point now worth to be called the reader atteention for is that care should be taken in applying straightforwardly the Feynman path integral eq. (1.50) to represent the propagator of a particle possessing fermionic degrees in the presence of an external Gauge field ([3]). One can avoid this operational path integral procedure by squaring the fermionic determinant and making use now of the well-defined proper-time formalism for bosonic coloured particles ([3]). Namely (see eq. (1.51))

$$
\begin{aligned}
& \det{}^{\frac{1}{2}}\big(\not{D}(igA+J)\ \not{D}(igA+J)\big) \\
&= \det{}^{\frac{1}{2}}\big(D^*(igA+J)D(igA+J)\big) \\
&\times \det\left(\mathbf{1} - \frac{ig}{4}[\gamma^\mu, \gamma^\nu](D^*D)^{-1}(igA+J) \times F_{\mu\nu}(igA+J)\right)
\end{aligned} \tag{1.64}
$$

Since the Klein-Gordon bosonic propagator can be written in term of the $SU(N)$ normalized holonomy factor as of as

$$
\begin{aligned}
& \big(D^*(igA_\mu + J_\mu)D(igA_\mu + J_\mu)\big)(x_1, x_2) \\
&= N \left\{ \int_0^\infty dt \int_{X(0)=x_1}^{X(t)=x_2} D[X(\sigma)]\, e^{-\frac{1}{2}\int_0^t \dot{X}^2(\sigma)\, d\sigma} \right\} \\
&\times \psi_{x_1 x_2}[C, A] \times \Phi_{x_1 x_2}[C, J]
\end{aligned} \tag{1.65}
$$

Here the non-abelian dynamical and abelian vectorial external sources phase factors are defined explicitly by

$$\psi_{x_1x_2}[C,A] = \left[\frac{1}{N}\mathbb{P}\left(\exp ig\int_0^t A_\mu(X(\sigma))dX^\mu(\sigma)\right)\right]$$
$$\Phi_{x_1x_2}[C,J] = \left[\exp\left(\int_0^t J_\mu(X(\sigma))dX^\mu(\sigma)\right)\right] \tag{1.66}$$

The final expression for the generating functional eq. (1.51) at large N, is thus easily written in the proper-time formalism, before taking the Yang-Mills path integral average is

$$\begin{aligned}
&\ell g\left\{\det{}^{\frac{1}{2}}\left(\not{D}^*(igA+J)\,\not{D}(igA+J)\right)\right\} \\
&= -\frac{1}{N}\left\{\int_0^\infty \frac{dt}{t}\int d^4x_1\int d\mu[C_{x_1x_2}]\right. \\
&\qquad \left.\left(\mathrm{Tr}_{SU(N)}\,\Psi_{x_1x_2}[C,A]\right)\Phi_{x_1x_2}[C,J]\right\} \\
&-\left\{\sum_{n=2}^{\infty}\left(\frac{1}{2}\right)^{n-1}\left(\frac{1}{1/N}\right)^n\right. \\
&\int d^4x_1\ldots d^4x_n\int_0^\infty (dt_1\ldots dt_n)\int d\mu[C_{x_1x_2}] \\
&d\mu[C_{x_nx_1}]\times \mathrm{Tr}_{\mathrm{Dirac}}\left([\gamma_{\mu_1},\gamma_{\nu_1}]\ldots[\gamma_{\mu_n},\gamma_{\nu_n}]\right) \\
&\times \mathrm{Tr}_{SU(N)}\left\{\frac{\delta}{\delta\sigma_{\mu_1\nu_1}(x_1)}\left(\psi_{x_1x_2}[C,A]\Phi_{x_1x_2}[C,A]\right)\right. \\
&\qquad \left.\left.\cdots\frac{\delta}{\delta\sigma_{\mu_n\nu_n}(x_n)}\left(\psi_{x_nx_1}[C,A]\Phi_{x_nx_1}[C,A]\right)\right\}\right\}
\end{aligned} \tag{1.67}$$

Here the Migdal-Makeenko loop derivative is introduced ([3])

$$\frac{\delta}{\delta\sigma_{\mu\nu}(X(\overline{\sigma}))} = \lim_{\varepsilon\to 0}\int_{-\varepsilon}^{\varepsilon} d\zeta.\zeta\frac{\delta^2}{\delta X_\mu(\overline{\sigma}+\frac{\zeta}{2})\delta X_\nu(\overline{\sigma}-\frac{\zeta}{2})} \tag{1.68}$$

which by its turn has the geometrical meaning of dividing the path trajectory $C_{x_1x_1}$, quite closely analogous to the joining and splitting picture of the old theory of dual strings (after taking the $SU(\infty)$ limit into our constant gauge fields model as given by eq. (1.31) or eq. (1.35) in section 3 of this chapter).

Another point worth to call attention is the expansion parameter on eq. (1.67) is the color N, but appearing now as a Laurent power series on $\left(\frac{1}{1/N}\right)^{+n}$.

Conclusion: We can see from this work that another time in QCD physics it is raised hopes that on underlying string dynamics is in place to handle correctly the mathematical – calculational aspects of Euclidean Non-Abelian Gauge – abelian Gauge theories in theirs confining phase, signaled here by a hypothesized non-vanishing energy for the non-perturbative vacuum $\big(\langle 0|tr(F^2)|0\rangle \neq 0\big)$. (See eq. (1.27) for the analytic expression of this hypothesis).

At this point let us remark that our string representation for the QCD-Eguchi-Kawai reduced model is a free bosonic one. However if one considers next non-constant full space-time variable corrections/fluctuations to the gauge connections entering into the full Yang-Mills path integrals, one is lead to the self-avoiding fermionic full structure of the $QCD(SU(\infty))$ ([3]) with the extrinsic string as an effective bosonic string representation for $QCD(SU(\infty))$ (chapter 3).

Finally, we should roughly say that our path integral is at $SU(\infty)$, but surely we are in the context of a somewhat $\frac{1}{D}$ expansion for the pure quantum Yang-Mills field, with a non perturbative vacuum. Unfortunately, the famous $\frac{1}{D}$ expansion of Lattice QCD has not been generalized or even well-understood on the continuum. We hope that our work should be a step in this direction.

1.5 Appendix A

Let us consider the term

$$J_1 = \exp\left\{\frac{1}{4}\, G_0^a\, G_1^b\, G_0^c\, G_1^d\, \mathcal{L}_{abcd} \times \left[g^2 V + \frac{(g^2 S)^2}{(N/2)}\right]\right\} \tag{A.1}$$

After the re-scaling

$$G_{0,1}^f = \widetilde{G}_{0,1}^f \left[g^2 V + \frac{(g^2 S)^2}{(N/2)}\right]^{-\frac{1}{4}} \tag{A.2}$$

It terms out to be

$$I_1 = \exp\left\{\frac{1}{4}\, \widetilde{G}_0^a\, \widetilde{G}_1^b\, \widetilde{G}_0^c\, \widetilde{G}_1^d\, \mathcal{L}_{abcd}\right\} \tag{A.3}$$

However a "mixed" term of the form

$$I_2 = \exp\left\{\frac{1}{4}\, G_0^a\, G_2^b\, G_0^c\, G_2^d\, \mathcal{L}_{abcd}(g^2 V)\right\} \tag{A.4}$$

under the re-scaling

$$G_2^f = \widetilde{G}_2^f \left[g^2 V\right]^{-\frac{1}{4}} \tag{A.5}$$

becomes now

$$I_2 = \exp\left\{\frac{1}{4}\,\widetilde{G}_0^a\,\widetilde{G}_2^b\,\widetilde{G}_0^c\,\widetilde{G}_2^d\,\mathcal{L}_{abcd}\,\frac{(g^2V)\times(g^2V)^{-1/2}}{\left[g^2V+\frac{(g^2s)^2}{N/2}\right]}\right\} \tag{A.6}$$

Note that $N\to\infty$, we have the leading asymptotic limit

$$\frac{(g^2V)^{1/2}}{\left[g^2V+\frac{(g^2S)^2}{N/2}\right]^{1/2}}\sim\frac{(g^2V)^{1/2}}{(g^2V)^{1/2}}\to 1. \tag{A.7}$$

It is worth recall that

$$\prod_{a=1}^{N^2-N}\left(dG_0^a\,dG_1^a\right)=\left\{\left[g^2V+\frac{(g^2S)^2}{N/2}\right]^{-\frac{(N^2-N)}{2}}\right\}\left(\prod_{a=1}^{N^2-N}d\widetilde{G}_0^a\,d\widetilde{G}_1^a\right) \tag{A.8}$$

$$\prod_{\substack{a=1\\ \mu\neq 0,1}}^{N^2-A}dG_\mu^a=\left\{(g^2V)^{-\frac{(N^2-N)}{4}(D-2)}\right\}\times\left(\prod_{\substack{a=1\\ \mu\neq 0,1}}^{N^2-N}d\widetilde{G}_\mu^a\right) \tag{A.9}$$

We added also the remark about the constant gauge field non-abelian Stokes theorem

$$\begin{aligned}
W[C] &= \mathrm{Tr}_{SU(N)}\left(\mathbb{P}\left(e^{ig\oint_c A_\mu\,dX_\mu}\right)\right\} \\
&= \frac{1}{N}\,\mathrm{Tr}_{SU(N)}\left\{\mathbb{P}\left(e^{ig\,F_{\mu\nu}\left(\int ds^{\mu\nu}\right)}\right)\right\} \\
&= \frac{1}{N}\,\mathrm{Tr}_{SU(N)}\left\{\mathbb{P}\left(e^{(ig)ig[A_\mu,A_\nu]_-\,S}\right\}\right\}(\delta_{\mu 0}\,\delta_{\nu 1}) \\
&\overset{N\to\infty}{\sim} \frac{1}{N}\,\mathrm{Tr}_{SU(N)}\left\{\mathbf{1}-g^2S[A_0,A_1]_-+\frac{(g^2S)^2}{2}\,[A_0,A_1]_-^2+\dots\right\} \\
&\overset{N\to\infty}{\sim} \exp\left\{+\frac{(g^2S)^2}{2N}\,\mathrm{Tr}_{SU(\infty)}\left([A_0,A_1]_-^2\right)\right\}
\end{aligned} \tag{A.10}$$

As a last point of our Wilson loop evaluations at large N at the context of constant gauge field configurations, we point out that at $D=2$ (the two dimensional case, it is not need to consider the phenomena of the dimensional transmutation coupling constant and the evaluation above displayed leads directly to the area behaviour for the Wilson Loop. (Remark originally due Bollini-Giambiagi). However it is important to keep in mind that such result can be obtained quite straightforwardly by using the axial gauge $A_0^a=0$, and mostly important, it shows the non-dynamical behaviour of the pure Yang-Mills quantum (perturbative) theory at two-dimensions. At

$D = 3$, our $SU(\infty)$-constant gauge field model yields charge screening instead of color charge confinement (a lenght behaviour for the Wilson Loop). Finally for $D > 4$, we have a infinite volume vanishing Wilson Loop, which by its term signals that the Yang-Mills theory in R^D, $D > 4$ is a trivial QFT, in place of the usual wrong, but always argued non-renormalizability of Yang-Mills theory for $D > 4$.

1.6 References

[1] Luiz C.L. Botelho - "Critical String Wave Equations and the $QCD(U(N_c))$ String. (Some comments)", Int. J. Theor. Phys. (2009) 48 - 2715-2725.

[2] Luiz C.L. Botelho - "The Electric Charge Confining in Polyakov's Compact QED in R^4", Int. J. Theor. Phys. (2009) 48 - 1695-1700.

[3] Luiz C.L. Botelho - "Methods of Bosonic and Fermionic Path Integral Representations - Continuous Random Geometry in Quantum Field Theory", Nova Science, New York - (2008).

[4] T. Eguchi and H. Kawai - "Reduction of Dynamical Degrees of Freedom in the Large N Gauge Theory", Phys. Rev. Lett., vol. 48, 1063 (1982).

[5] Luiz C.L. Botelho - "The confining behavior and asymptotic freedom for $QCD(SU(\infty))$ - a constant Gauge field path integral analysis", Eur. Phys. J. 44, 267-276 (2005).

[6] Luiz C.L. Botelho - "Triviality - Quantum Decoherence of Fermionic Chronodynamics $SU(N_c)$ in the Presence of an External Strong $U(\infty)$ Flavored Constant Noise Field", Int. J. Theor. Phys. (2010) 49 - 1684-1692.

[7] M.N. Chernodubat all - "On chromoeletric (super) conductivity of the Yang-Mills vacuum", avXiv: 1212.3168 v 1 [hep-ph], 13 Dec. 2012.

[8] Claude Itzykson & Jean-Michel Drouffe - "Statistical field theory", vol. 2, Cambridge Monographs on Mathematical Phisics, 1991.

- Luiz C.L. Botelho - Research Article - "Basics Polyakov's Quantum Surface Theory on the Formalism of Functional Integrals and Applications", International Scholarly Research Network ISRN High Energy Physics, vol. 2012, Article ID674985, doi: 10.5402/2012/674985.

[9] N.N. Bogoluhov, A.A. Logunov and I.T. Todorov - Introduction to Axiomatic Quantum Field Theory - Mathematical Physics Monograph series - Benjamin/Cummings Publishing Company Inc. - USA, 1978.

Chapter 2

Basics Polyakov's quantum surface theory on the formalism of functional integrals and applications

2.1 Introduction

Since its inception on seventy years ago, Non-Abelian Gauge theories have shown as the most promissing mathematical formalism for a realistic description of strong interactives and even formulated on its supersymmetric version it has became an attractive attempt for unify Physics.

In strong interaction the picture image of a mesonic quantum excitation, is for instance, a wave quantum mechanical functional assigned to a classical configuration of a space-time trajectory of a pair quark-antiquark bounding a space-time non-abelian gluon flux surface (of all topological genera) connecting both particle pair: the famous t'Hoft-Feymman planar diagrams ([1]).

It appears thus, appealing for mathematical formulations to consider directly as dynamical variables or wave functions in this Faraday line framework for non-abelian gauge theories, the famous quantum Wilson Loop or (quantum) holonomy factor associated to a given space-time Feynman quark-antiquark closed trajectory C in a $SU(N_c)$ Yang-Mills quantum field theory. Namely

$$\psi[C] = \langle W[C] \rangle = \left\langle \frac{1}{N_c} \left\{ \mathrm{Tr}^{(c)}_{\mathrm{SU(N_c)}} \left[\mathbb{P} \left(\exp \mathrm{ig} \oint_{\mathrm{C}} \mathrm{A}_\mu \, \mathrm{dx}_\mu \right) \right] \right\} \right\rangle \tag{2.1}$$

Here $\langle \ \rangle$ is defined by the Yang-Mills quantum path-integral and $A_\mu(x) = A^c_\mu(x)\lambda_a$ denotes the quantized Yang-Mills (Gluon) connection.

It is thus, searched loop space dynamical equations (at least on the formal level without considering those famous ultra-violet "renormalization problems") for the new wave function Eq. (2.1). However Lattice Yang-Mills theories has indicated that "string solutions" with a symbolic structure form

$$\psi[C] = \sum_{S,\partial s=C} \left\{ \left[\exp\left(-\frac{1}{2\pi\alpha'} A(S) \right) \right] \Phi^{QCD}[C] \right\} \tag{2.2}$$

with $\sum_{S;\partial S=C}$ denoting the Feynman-Wiener continuous sum over all surfaces S, with all topological genera and bounded by C; α' is the fundamental constant of strong force physics called the Regge slope parameter ([1]) $A(S)$ is the area functional "evaluated" at the sample surface S and $\Phi[C]$ is a functional related probably to the existence of a (neutral) non-abelian 2D intrinsic fermions structure on the surface S, called the Elfin fermionic functional ([1]).

It is a consequence of eq. (2.1)-eq. (21.2) that non-abelian gauge theories (even on theirs supersymmetric versions ([2])) should be better reformulated as a dynamics of random surfaces (strings theories).

Our aim in this chapter is to present in full technical details added with ours original improvements, the work of A.M. Polyakov ([1]) to give a precise path integral meaning for eq. (2.2) in a 2D-gravitational context and only considering the case of trivial surface topology.

We give (also in details) the path integral meaning for the usual case of Nambu-Goto for eq. (2.2) in the pure surface context, all these new results due to ourselves ([1]).

This chapter is organized as follows: in §1.2 we survey some basic results of classical surface theory. In §1.3, we expose the Physical toy model exactly soluble Polyakov's framework. In §1.4, we expose the Nambu-Goto theory (our results).

2.2 Elementary results on the Classical (Bosonic) Surface theory

A given surface S, with a boundary being a curve C and embedded into a euclidean space-time R^D is usually described by a vector field $X_\mu(\xi_1, \xi_2)$ ($\mu = 1, \ldots, D$) and a given two-dimensional domain Ω, compatible with the topology of S in the case of the surface's hypothesis smoothness. It is imposed also that such vector field, called from now on as surface vector position field, when restricts to the boundary of Ω should be a reparametrization of C $(X_\mu(\xi_1, \xi_2))_{\partial\Omega} = C)$.

The Nambu-Goto area functional associated to a given surface vector position $(X_\mu(\xi_1, \xi_2))$ is given by usual Riemann integral $((\xi_1, \xi_2) := \xi)$.

$$A(S_C) = \int_\Omega \left(\det\{h_{ab}(\xi)\} \right)^{1/2} d^2\xi \tag{2.3}$$

with $h_{ab}(\xi) = (\partial_a X^\mu \partial_b X_\mu)(\xi)$ denoting the metric tensor induced on the surface S_c (rigorously induced on the surface-manifold tangent bundle) by the parametrization $\{X_\mu(\xi)\}$.

The most important property of the functional eq. (2.3) is its invariance under the (formal) group of the reparametrizations of S. Namely:

$$\begin{aligned}(\xi_1, \xi_2) &= (\varphi_1(\xi_1', \xi_2'); \varphi_2(\xi_1', \xi_2') \\ X_\mu'(\xi') &= X_\mu(\varphi(\xi_1'))\end{aligned} \tag{2.4}$$

here $\varphi(\xi')$ denotes a two-dimensional intrinsic C_1 vector field on Ω with everywhere non-zero Jacobian.

Formal Euler-Lagrange equations associated to the surface action functional eq. (2.3) can be easily written and producing the boundary value problem on Ω for each μ-component ($\mu = 1, \ldots, D$)

$$\left| \begin{aligned} \Delta_h X^\mu(\xi) &= 0 \\ X^\mu(\xi))\big|_{\xi \in \partial\Omega} &= C^\mu \end{aligned} \right. \tag{2.5}$$

where Δ_h denotes the second order elliptic operator called Laplace-Beltrami associated to the metric $(h_{ab}(\xi))$ $(h(\xi) = \det\{h_{ab}(\xi)\})$

$$\Delta_h = \left(\frac{1}{\sqrt{h}} \partial_a (\sqrt{h}\, h^{ab}\, d_b) \right)(\xi) \tag{2.6}$$

If one now choose the conformal gauge for the surface $h_{ab}(\xi) = e^{\varphi(\xi)} \delta_{ab}$, formally one obtains that the surface vector position satisfies the Dirichlet problem in Ω.

$$\left| \begin{aligned} \Delta_{h=\delta_{ab}} X^\mu(\xi) &= 0 \\ X^\mu(\xi))\big|_{\xi \in \partial D} &= C^\mu \end{aligned} \right. \tag{2.7}$$

The solution of the above mentioned potential problem can be exactly given by conformal complex variable theory methods ([2] - Chapter 1).

Note that eq. (2.7) produces minimal surfaces (i.e. classical surfaces which minimize the area functional locally) as the associated classical surfaces actions on the Nambu-Goto theory eq. (2.3). Note that all the above displayed discussion remains correct to the case of general topological genera (i.e. $\chi(S) = \frac{1}{4\pi} \int_\Omega (\sqrt{h}\, R(h))(\xi)\, d\xi^2 \neq 0$; here $R(h)$ denoting the scalar of curvature associated to the induced metric $h_{ab}(\xi)$).

However at the classical level, there is a quadratic area functional due to Howe-Brink-Polyakov equivalent to the above result related to the classical aspects Nambu-Goto action. It is the functional associated to a theory of D scalar classical massless fields on Ω but interacting with intrinsic (news) dynamical degrees of freedom given by the infinite-dimensional manifold of two-dimensional metrical structures on Ω.

The classically equivalent area action functional is now given by the massless fields $\{X^\mu(\xi)\}$ gravitation content

$$\overline{A}(S_C) := \frac{1}{2}\int_\Omega (\sqrt{g}\, g^{ab}\, \partial_a X^\mu\, \partial_b X_\mu)(\xi) d^2\xi \tag{2.8}$$

where $\{g_{ab}(\xi)\}_{\substack{a=1,2\\ b=1,2}}$ belongs to the infinite-dimensional space-manifold of all possible metrics admissible by the surface S – the famous exactly soluble string Polyakov' model.

That more rich (from a dynamically point of view) surface action functional is now invariant (formally) under the extended group of surface reparametrizations (the group of local diffeomorphism of the surface S).

$$\begin{aligned}
\xi_a' - \xi_a &= \delta\xi_b = \mathcal{E}_a(\xi)\\
\delta X_\mu(\xi) &= \mathcal{E}_a(\xi)\partial^a X_\mu(\xi)\\
\delta g_{ab}(\xi) &= (\nabla_a\, \mathcal{E}_b + \nabla_b\, \mathcal{E}_a)(\xi).
\end{aligned} \tag{2.9}$$

Here the surface's reparametrization vector fields $\{\mathcal{E}_a(\xi)\}$ are such that $(\nabla^a\, \mathcal{E}_a)(\xi) = 0$, a necessasry restriction in order to take out from the above written reparametrizations, all those which act on as a simply metrical rescalings. Note that we have imposed further the vector field $\mathcal{E}_a(\xi)$ vanishing at the boundary of S (i.e. $\mathcal{E}_a(\xi)\big|_{\partial\Omega} \equiv 0$). The covariant derivative is usually defined by the Christoffel-Schwarzt objects associated to the given metric tensor $\{g_{ab}(\xi)\}$.

$$(\nabla_a\, \mathcal{E}_b)(\xi) = (g_{aa'}\, \nabla^{a'})(\mathcal{E}_b)(\xi) = (\partial_a\, \mathcal{E}_b - \Gamma^c_{ab}\, \mathcal{E}_c)(\xi) \tag{2.10-a}$$

$$\Gamma^c_{ab}(\xi) = \frac{1}{2}\{g^{cd}(\partial_a\, g_{bd} + \partial_b\, g_{ad} - \partial_a\, g_{ab})\}(\xi) \tag{2.10-b}$$

In the important case of the metric field $g_{ab}(\xi)$ has the conformal form $g_{ab}(\xi) = e^{\varphi(\xi)}\,\delta_{ab}$, the above written objects take the simple forms below written

$$\begin{aligned}
\Gamma^c_{ab}(\xi) &= \frac{1}{2}\,\delta^{cd}\, e^{-\varphi}\big[\partial_a\varphi\, e^\varphi\, \delta_{bd} + \partial_b\varphi\, e^\varphi\delta - \partial_d\varphi\, e^\varphi\, \delta_{ab}\big]\\
&= \frac{1}{2}\,(\partial_a\varphi)\delta^{cb} + \frac{1}{2}(\partial_b\varphi)\delta^{ca} - \frac{1}{2}\,\partial_c\varphi\,\delta_{ab}
\end{aligned} \tag{2.11-a}$$

$$\nabla_a\mathcal{E}_b = \partial_a\mathcal{E}_b - \frac{1}{2}(\partial_c\varphi)\mathcal{E}^c\,\delta_{ab} - \frac{1}{2}\big[(\partial_a\varphi)\mathcal{E}_b - (\partial_b\varphi)\mathcal{E}_a\big] \tag{2.11-b}$$

Note that one has in addition to the usual reparametrization invariance, one further pivotal (point dependent) new symetry called the Weyl conformal symetry which acts solely on the new degree of freedom (with $\lambda(\xi) > 0$), crucial for the exactly modeel solubility.

$$g_{ab}(\xi) \to \lambda(\xi) g_{ab}(\xi). \tag{2.12}$$

The classical motion equations associated to the Howe-Brink-Polyakov functional are easily written down:

$$\Delta_{g_{ab}} X^\mu(\xi) = 0$$
$$X^\mu(\xi)\big|_{\partial\Omega} = C^\mu$$

$$T_{ab}(\xi) = (\partial_a X_\mu \partial_b X^\mu - \frac{1}{2} g_{ab} g^{cd} \partial_c X^\mu \partial_d X_\mu)(\xi) = 0. \tag{2.13}$$

From the last classical motion equation (2.13) for the intrinsic metric field, one obtains the result

$$g_{ab}(\xi) = \overline{\lambda}(\xi)(\partial_a X_\mu \partial_b X^\mu)(\xi) \tag{2.14}$$

where the unknow scale $\overline{\lambda}(\xi)$ can be formally fixed to be the function $\overline{\lambda}(\xi) = 1$ on the interior of Ω.

$$\big(g_{ab}(\xi)\big)_{\partial\Omega} = \big(g_{ab}(\xi)\big)_{\partial\Omega} = \delta_{ab}. \tag{2.15}$$

Another important classical surface functional, probably related to the still not completely understood functional $\Phi^{QCD}[C]$ on eq. (2.2) is the so called extrinsic functional which is defined by the square of the surface mean curvature (see eq. (2.6)))

$$(\mathcal{H}^?(S)) = (\Delta_h X^\mu)^2 = \left\{ -\frac{1}{\sqrt{h}} \partial_a(\sqrt{h}\, h^{ab} \partial_b)(X^\mu \right\}^2. \tag{2.16}$$

Namely

$$\mathcal{F}_{extr}(S) = \int_\Omega \sqrt{h}\,((\Delta_h X^\mu)^2(\xi))\, d^2\xi. \tag{2.17}$$

In the Polyakov's version of two-dimensional Ω quantum gravity, the above functional can be replaced by the fourth-order scalar field action

$$\begin{aligned} \mathcal{F}_{ext}[X^\mu(\xi), g_{ab}(\xi)] = &\int_\Omega (\sqrt{g}(\Delta_g X^\mu)^2(\xi)) d^2\xi \\ &+ \int_\Omega (\sqrt{g}\lambda^{ab}(\xi)(g_{ab}(\xi) - h_{ab}(\xi, X^\mu)))(\xi) \end{aligned} \tag{2.18}$$

where $\lambda^{ab}(\xi)$ denote lagrange multipliers insuring that at the classical level $g_{ab}(\xi) = (\partial_a X^\mu \partial_b X_\mu)(\xi)$.

Boundary conditions to be imposed on the complete action $A(S) + \mathcal{F}_{ext}(S)$ are mathematical subtle in order to guarantee the mathematical good property of strong elliptic of the associated boundary value problem ([3] - Chapter 5).

For instance, if one consider the case $g_{ab} = \delta_{ab}$ and the associated biharmonic operator

$$(\Delta X^\mu)^2 = \frac{\partial^4 X^\mu}{\partial \xi_1^4} + 2\frac{\partial^4 X^\mu}{\partial \xi_1^2 \partial \xi_2^2} + \frac{\partial^4 X^\mu}{\partial \xi_2^4}, \tag{2.19-a}$$

one can associate the following system of boundary operators

$$\begin{aligned} X^\mu(\xi)\big|_{\partial\Omega} &= C^\mu(\xi) \\ -\frac{\partial}{\partial n} X^\mu(\xi)\big|_{\partial\Omega} &= C^\mu(\xi). \end{aligned} \tag{2.19-b}$$

Here the meaning of C^μ is not so clear as an independent dynamical parameter of the loop C (string).

In the following discussions we will always regard the fourth-order actions as effective actions coming from the integrating out fermionic intrinsic degrees of freedom ([1]).

We left to our readers to prove the Green's formula for the fourth-order problem eq. (2.19)

$$\begin{aligned} \int_\Omega (f\,\Delta^2 g) d^2\xi &= \int_\Omega (g\,\Delta^2 f) d^2\xi \\ &+ \int_{\partial\Omega} \left(-f\,\partial\frac{\Delta g}{\partial n} + \frac{\partial f}{\partial n}\Delta g + g\frac{\partial}{\partial n}\Delta f - \frac{\partial g}{\partial n}\Delta f \right) ds. \end{aligned} \tag{2.20}$$

As a further comment and just for the reader's curiosity, one has an analogue of representing (at least locally) two-dimensional harmonic functions on Ω by analytical complex variable functions; one still has the following representation (not unique!) for biharmonic functions (the Goursat representation - 1898)

$$\mathcal{V}(\xi_1, \xi_2) = \mathrm{Re}\{\overline{z}\,\varphi(z) + \rho(z)\} \tag{2.21}$$

where $z = \xi_1 + i\xi_2$ and $(\varphi(z), \rho(z))$ a pair of complex variable analytic functions on Ω.

If the surface S is on R^3 and given in "nonparametric form" $X^3 = f(X^2, X^2)$, we can introduce as parametric coordinates $\xi^1 = X^1$, $\xi^2 = X^2$ and obtain as a result convenient for computations (exercises for the diligent reader), the Gauss curvature given explicitly by

$$R(h_{ab}) = \left(1 + f_{X^1}^2 + f_{X^2}^2\right)^2 \cdot \det\{f_{ab}(X^1, X^2)\}. \tag{2.22-a}$$

And for the mean curvature, the following result:

$$H = \left(1 + f_{X^1}^2 + f_{X^2}^2\right)^{-3/2} \times \left\{\left(1 + f_{X_2}^2\right) f_{X_1X_1} - 2\, f_{X_1}\, f_{X_2}\, f_{X_1X_2} + \left(1 + f_{X_1}^2\right) f_{X_2X_2}\right\} \quad \text{(2.22-b)}$$

It is worth recall that in the important case of a surface S embedded in R^3 (useful in Surface's Statistical Physics), but with the general parametrization:

$$X^1 = X^1(\xi_1, \xi_2); \quad X^2 = X^2(\xi_1, \xi_2); \quad X^3 = X^3(\xi_1, \xi_2);$$

the expression for the area functional and the surface extrinsic functional can be given explicitly by (exercise for our diligent readers!)

$$A(S) = \int_\Omega \left(\sqrt{EG - F^2}\right)(\xi)\, d^2\xi \quad \text{(2.23-a)}$$

$$\mathcal{F}_{ext}(S) = \int_\Omega H^2(\xi)\, d^2\xi \quad \text{(2.23-b)}$$

where the surface objects are given by

$$E = \left(\frac{\partial X^\mu}{\partial \xi_1}\right), \left(\frac{\partial X_\mu}{\partial \xi_1}\right); \quad F = \left(\frac{\partial X^\mu}{\partial \xi_1}\right), \left(\frac{\partial X_\mu}{\partial \xi_2}\right);$$

$$G = \left(\frac{\partial X^\mu}{\partial \xi_2}\right), \left(\frac{\partial X_\mu}{\partial \xi_2}\right);$$

$$H = \frac{1}{(EG - F^2)}(ED'' - 2D'F + DG);$$

$$D = \frac{1}{\sqrt{EG - F^2}} \begin{vmatrix} X^1_{\xi_1\xi_1} & X^2_{\xi_1\xi_1} & X^3_{\xi_1\xi_1} \\ X^1_{\xi_1} & X^2_{\xi_1} & X^3_{\xi_1} \\ X^1_{\xi_2} & X^2_{\xi_2} & X^3_{\xi_2} \end{vmatrix}$$

$$D' = \frac{1}{\sqrt{EG - F^2}} \begin{vmatrix} X^1_{\xi_1\xi_2} & X^2_{\xi_1\xi_2} & X^3_{\xi_1\xi_2} \\ X^1_{\xi_1} & X^2_{\xi_1} & X^3_{\xi_1} \\ X^1_{\xi_2} & X^2_{\xi_2} & X^3_{\xi_2} \end{vmatrix} \quad \text{(2.23-c)}$$

$$D'' = \frac{1}{\sqrt{EG - F^2}} \begin{vmatrix} X^1_{\xi_2\xi_2} & X^2_{\xi_2\xi_2} & X^3_{\xi_2\xi_2} \\ X^1_{\xi_1} & X^2_{\xi_1} & X^3_{\xi_1} \\ X^1_{\xi_2} & X^2_{\xi_2} & X^3_{\xi_2} \end{vmatrix}$$

Finally we remark that one could easily generalize all the above written results to the general case of the initial ambient space R^D being new a general Riemann Manifold $(M, G_{\mu\nu}(X^\gamma))$. In this case, the "induced" surface metric tensor $h_{ab}(X(\xi)) = \partial_a X^\mu(\xi)\partial_b X_\nu(\xi)$ will be replaced by the new "induced" tensor

$$h_{ab}(X(\xi), [G_{\mu\nu}]) = \partial_a X^\mu(\xi) G_{\mu\nu}(X^\gamma(\xi))\partial_b X^\nu(\xi)$$

on the exposed formulae. The full classical and quantum theory of there surfaces σ-models can be found in [1].

2.3 Path integral quantization on Polyakov's theory of surface (or how to quantize 2D massless scalar fields in the presence of 2D quantum gravity)

After exposing some basic concepts of the classical surface theory in Section 1.2, we now pass to the vital problem of quantization on the formalism of Feynman-Wiener path integrals.

Following R.P. Feynman in his theory of path integration – sum over histories, a mathematical meaning for the continuous sum over (now), euclidean quantum (random) surfaces for eq. (1.2) should be given by following path integrals for the Nambu-Goto case and of A.M. Polyakov respectively (see eqs. (2.3)–(2.8)).

$$G_{NG}(C) = \int_{X_\mu(\xi)\big|_{\xi\in\partial\Omega}=C_\mu} d\mu[X^\mu(\xi)] \exp\left\{-\frac{1}{2\pi\alpha'}A(S_c)\right\} \tag{2.24-a}$$

$$G_p(C) = \int_{\substack{X_\mu(\xi)\big|_{\xi\in\partial\Omega}=C_\mu \\ g_{ab}(\xi)\big|_{\xi\in\partial\Omega}=0}} d\overline{\mu}[X^\mu(\xi), g_{ab}(\xi)] \exp\left\{-\frac{1}{2\pi\alpha'}\overline{A}(S_c)\right\} \times \exp\left\{-\mu_0^2\int_\Omega \sqrt{g}(\xi)\, d^2\xi\right\} \tag{2.24-b}$$

Here α' and μ_0^2 are constants to be identified with certain physical parameters of the theory. Note that α' has dimension of inverse if mass square. And $[\mu_0^2] = [\alpha']^{-1}$. It is important remark that the Feynman-Wiener path measures $d\mu[X^\mu(\xi)]$ and $d\overline{\mu}[X^\mu(\xi), g_{ab}(\xi)]$ must be defined in a such way in order to be fully invariant under the action of the local diffeomorphism group of both descriptions (see eq. (2.4) and eq. (2.9)).

We aim now to evaluate explicitly the A.M. Polyakov's integral eq. (1.24-b). As a first step one recall that we have imposed the "fluctuating" zero Dirichlet boundary condition for the intrinsic metric field $g_{ab}(\xi)\big(g_{ab}(\xi)\big|_{\partial\Omega} = 0\big)$. That results lead us to the effective action for the surface dynamics weight functional

$$\overline{A}(S_c) = \frac{1}{2}\int_\Omega \big(\sqrt{g}\, X^\mu(-\Delta_g)X_\mu\big)(\xi)\, d^2\xi. \tag{2.24-c}$$

Since at the classical level, the metric field decouples (one can choose it in the form $g_{ab} = e^\varphi\, \delta_{ab}$) and considering the usual classical plus quantum decomposition $X_\mu(\xi) = X_\mu^{CL}(\xi) + \sqrt{(\pi\alpha')}\, X_\mu^q(\xi)$, where the quantum correction $X_\mu^q(\xi)$ is such that $X_\mu^q(\xi)\big|_{\partial\Omega} = 0$.

After these preliminaries considerations one is lead to evaluate the following covariant path integral (after disregarding classical field contributions to the path integral)

$$I[g_{ab}] = \int d\mu[X^\mu(\xi)] \exp\left\{-\frac{1}{2}\int_\Omega (\sqrt{g}\, X^\mu(-\Delta_g)X_\mu)\,(\xi)\, d^2\xi\right\}. \quad \text{(2.24-d)}$$

It is worth recall that there is a classical term (a functional depending on the loop boundary C) coming from the partial integration to arrive at eq. (2.24) for the case $g_{ab} = \delta_{ab}$, which clearly do not satisfy the quantum-fluctuating nature of these non-constant $g_{ab}(\xi)$

$$\text{Classical Term } = \int_\Omega (\partial_a X^\mu_{CL} \partial_a X^{CL}_\mu)(\xi) d^2\xi. \quad (2.25)$$

The Feynman-Wiener measure $d\mu[X^\mu(\xi))$ on eq. (2.24-d) must be choose in order to be invariant under the eq. (2.9). The only local candidate is given by the "weighted" product Feynman measure ([1])

$$d\mu[X^\mu(\xi)] := \prod_{\xi\in\Omega} \left(\sqrt[4]{g(\xi)}\, dX^\mu(\xi)\right) \quad (2.26)$$

As a consequence the path integral eq. (2.24) turns out to be Gaussian and metric dependent. It yields the immediate result:

$$I[g_{ab}] = \det{}^{-D/2}(-\Delta_g). \quad (2.27)$$

We thus, have reduced the "explicitly" evoluation of the Polyakov's path integral to the functional integration of the above written functional over all fluctuating metric fields

$$I = \int d\mu[g_{ab}] \det{}^{-D/2}(-\Delta_g) e^{-\mu_0^2 \int_\Omega \sqrt{g} d^2\xi}. \quad (2.28)$$

By using the theory of invariant integration on Riemann Manifolds of ours ([1]), one can insures automatically the preservation of the diffeomorphism group by defining the above mentioned metric $d\mu[g_{ab}]$ as the (functional) element of volume of the following functional metric (so called De Witt metric) with $(\neq -\frac{1}{2})$:

$$||\delta g_{ab}||^2 = \int_\Omega \left[\sqrt{g}\left\{(\delta g_{ab})\,(g^{aa'}\, g^{bb'} + c\, g^{ab}\, g^{a'b'})\,(\delta g_{a'b'})\right\}\right](\xi)\, d^2\xi \quad (2.29)$$

Let us choose the conformal gauge fixing to evaluate the above metric path integral sucessfully ([1]), but already considering the metric positivity

$$\overline{g}_{ab}(\xi) = e^{\varphi(\xi)}\, \delta_{ab}\,. \quad (2.30)$$

As a result the functional infinitesimal displacements $\delta g_{ab}(\xi)$ can be written in the following general form on the functional manifold of the metric $[g_{ab}(\xi)]$ (on the formal infinite-dimensional functional tangent bundle)

$$\delta g_{ab}(\xi) = \delta\varphi(\xi)\overline{g}_{ab}(\xi) + (\nabla_a \mathcal{E}_b + \nabla_b \mathcal{E}_a)(\xi). \tag{2.31}$$

Let us re-write the functional De-Witt metric in the form

$$||\delta g_{ab}||^2 = \int_\Omega (\delta g_{ab}(\xi))\,\gamma^{(ab,a'b')}\,[g_{cd}(\xi)](\delta g_{a'b'}(\xi)) \tag{2.32-a}$$

$$\gamma^{(ab,a'b')}\,[g_{cd}(\xi)] = \left(\sqrt{g}(g^{aa'}\,g^{bb'} + c\,g^{ab}\,g^{a'b'})\right)(\xi) \tag{2.32-b}$$

Since there is only three independent components of the metric tensor $g_{ab}(\xi)$ in two-dimensions ($g_{12} = g_{21}$), the metric coeficients $\{\overline{\gamma}_{ij}\}_{\substack{1\le i\le 3\\ 1\le j\le 3}}$ are effectively a 3×3 matrix on the functional manifold of the intrinsic 2D metric fields

$$||\delta g_{ab}||^2 = \int_\Omega \left\{ (\delta g_{11}, \delta g_{12}, \delta g_{22})[\overline{\gamma}] \begin{pmatrix} \delta\, g_{11} \\ \delta\, g_{12} \\ \delta\, g_{22} \end{pmatrix} \right\} (\xi)\, d^2\xi$$

$$= \int \Big\{ (\delta g_{11})(\overline{\gamma}_{11})(\delta g_{11}) + (\delta g_{12})(\overline{\gamma}_{22})(\delta g_{22})$$

$$+(\delta g_{22})(\overline{\gamma}_{33})(\delta g_{22}) - (\delta g_{11})(\overline{\gamma}_{12} + \overline{\gamma}_{21})(\delta g_{12})$$

$$+(\delta g_{11})(\overline{\gamma}_{13} + \overline{\gamma}_{31})(\delta g_{22}) + (\delta g_{12})(\overline{\gamma}_{23} + \overline{\gamma}_{32})(\delta g_{22}) \Big\} (\xi) \times d^2\xi \tag{2.33}$$

Here the explicitly expressions of the effective $[\overline{\gamma}]$-matrix is related to these of eq. (2.32-b) by the results

$$\begin{aligned}
&\overline{\gamma}_{11} = \gamma^{(11,11)} = e^{\varphi}(e^{-\varphi}\,\delta^{11}e^{-\varphi}\,\delta^{11} + ce^{-\varphi}\,\delta^{11}\,e^{-\varphi}\,\delta^{11}) = (1+c)e^{-\varphi};\\
&\overline{\gamma}_{12} = \gamma^{(11,12)} = 0; \quad \overline{\gamma}_{21} = \gamma^{(12,11)} = 0; \quad \overline{\gamma}_{13} = \gamma^{(11,22)} = c\,e^{-\varphi}\\
&\overline{\gamma}_{31} = \gamma^{(22,11)} = c\,e^{-\varphi}; \quad \overline{\gamma}_{23} = \gamma^{(12,22)}; \quad \overline{\gamma}_{33} = \gamma^{(22,12)} = (1+c)e^{-\varphi}
\end{aligned} \tag{2.34}$$

As a consequence

$$\det[\gamma^{(ab,a'b')}] = \det[\overline{\gamma}_{i\delta}] = \frac{(1+c)^2}{e^{3\varphi}} - \frac{c^2}{e^{3\varphi}} = (1+x)e^{-3\varphi}. \tag{2.35}$$

So the De Witt metric on 2D is non-singular only if ($C \neq -\frac{1}{2}$).

Let us substitute the general funcitonal displacement eq. (1.31) on eq. (2.3-c).

Firstly we get (one could choose from the begining on eq. (2.32-b).

$$||\delta g_{ab}||^2 = \int_\Omega d^2\xi\, e^{\varphi(\xi)}\{(\delta\varphi g_b^a + \nabla^a \mathcal{E}_b + \nabla_b \mathcal{E}^a) \cdot (\delta\varphi g_a^b) + \nabla^b \mathcal{E}_a + \nabla_a \mathcal{E}^b) + 4c(\nabla_c \mathcal{E}^c + \delta\varphi)^2\} \tag{2.36}$$

Let us analyze those terms involving the conformal factor.

By noting that $\nabla_c g_{ab} = \nabla_c g^{ab} = 0$ and $(\nabla_a \mathcal{E}_b) g^{aa'} g^{bb'} (\nabla_{b'} \mathcal{E}_{a'}) = \nabla^{a'} \mathcal{E}_b \nabla^b \mathcal{E}_{a'}$, added with $\nabla_a \mathcal{E}^a = 0$; $\nabla_{a'}(g^{bb'} \mathcal{E}_{b'}) = g^{bb'} \nabla_{a'} \mathcal{E}_{b'} = \nabla_{a'} \mathcal{E}^b$; one obtains the chaim of results below written

$$\begin{aligned}
||\delta g_{ab}||^2 &= \int_\Omega d^2\xi\, e^{\varphi(\xi)}\{(\delta\varphi g_b^a + \nabla^a \mathcal{E}_b + \nabla_b \mathcal{E}^a) \\
&\quad \cdot (\delta\varphi g_a^b) + \nabla^b \mathcal{E}_a + \nabla_a \mathcal{E}^b) + 4c(\nabla_c \mathcal{E}^c + \delta\varphi)^2\}(\xi) \\
&= \int_\Omega d^2\xi\, e^{\varphi(\xi)}\{(\nabla^a \mathcal{E}_b + \nabla_b \mathcal{E}^a)(\nabla^b \mathcal{E}_a + \nabla_a \mathcal{E}^b) \\
&\quad - 2(\nabla_d \mathcal{E}^d)^2 + 2(1+2c)(\nabla_d \mathcal{E}^d + \delta\varphi)^2\}(\xi) \\
&= 2(1+2c)\int_\Omega d^2\xi\, e^{\varphi(\xi)}(\nabla_d \mathcal{E}^d + \delta\varphi)^2(\xi) \\
&\quad - 2\int_\Omega d^2\xi\, e^{\varphi(\xi)}\{\mathcal{E}^a(\nabla_d \nabla^d + [\nabla^b, \nabla_a])\mathcal{E}_b\}
\end{aligned} \tag{2.37}$$

where we have used the covariant by parts integration rule $(g|_{\partial\Omega} = 0)$

$$\int_\Omega d^2\xi\, \sqrt{g}\, f(\nabla_d h) = -\int_\Omega d^2\xi\, \sqrt{g}(\nabla_d f)h. \tag{2.38}$$

As a consequence we have the full result

$$\begin{aligned}
||\delta g_{ab}||^2 &= (1+2c)\int_\Omega d^2\xi\, \sqrt{g(\xi)}(\delta\mathcal{V}(\xi))^2 \\
&\quad + \int_\Omega d^2\xi\, \sqrt{g(\xi)} \left|(\mathcal{E}_1(\xi), \mathcal{E}_2(\xi))[\widehat{\mathcal{L}}]\begin{pmatrix}\mathcal{E}_1(\xi)\\ \mathcal{E}_2(\xi)\end{pmatrix}\right|
\end{aligned} \tag{2.39}$$

where

$$\delta\mathcal{V}(\xi) = \delta\varphi(\xi) + \overbrace{(\nabla^c \mathcal{E}_c)(\xi)}^{=0} = \delta\varphi(\xi) \tag{2.40}$$

and the Faddev-Popov operator acting on the two-dimensional vector field generators of the local group of coordinate transformations is explicitly given by

$$(\widehat{\mathcal{L}}\mathcal{E})_a = -((\nabla^d \nabla_d \delta_a^b + [\nabla^b, \nabla_a])\mathcal{E}_b)(\xi) \tag{2.41}$$

As a consequence the functional volume element writes as of as

$$d\mu[g_{ab}(\xi)] = d^{cov}[\varphi(\xi)]d^{cov}[\mathcal{E}_1(\xi)]d^{cov}[\mathcal{E}_2(\xi)]\det{}^{1/2}[\mathcal{L}]. \tag{2.42}$$

Here

$$\begin{aligned} d^{cov}(\varphi(\xi)) &= \prod_{(\xi\in\Omega)} e^{\frac{\varphi(\xi)}{2}}\,\delta\varphi(\xi) = \prod_{\xi\in\Omega}\delta\left(2\,e^{\frac{\varphi(\xi)}{2}}\right) = D^F\left[e^{\frac{\varphi(\xi)}{2}}\right] \\ d^{cov}[\mathcal{E}_a(\xi)] &= \prod_{\xi\in\Omega} d\,\mathcal{E}_a(\xi) = D^F[\mathcal{E}_a(\xi)] \end{aligned} \tag{2.43}$$

Just for pedagogical purpose let us evaluate in details the relevant term for the Fadeev-Popov operator in the above formulae.

We have that, for instance, the following sample calculation

$$\begin{aligned} &\int_\Omega d^2\xi\,\sqrt{g}(\nabla_a\mathcal{E}_b + \nabla_b\mathcal{E}_a)g^{aa'}\,g^{bb'}(\nabla_{a'}\mathcal{E}_{b'} + \nabla_{b'}\mathcal{E}_a) \\ &= \left(\int d^2\xi\{\sqrt{g}(\nabla_a\mathcal{E}_b)g^{aa'}\,g^{bb'}(\nabla_{a'}\mathcal{E}_{b'})\}(\xi)\right) + \text{similar terms} \\ &= \int d^2\xi\sqrt{g}\Big[\overbrace{(\nabla^a\mathcal{E}_b,\nabla_a\mathcal{E}^b)}^{g^{bb'}\nabla_{a'}\mathcal{E}_{b'}=\nabla_{a'}(g^{bb'}\mathcal{E}_{b'})=\nabla_{a'}\mathcal{E}^b} \\ &\qquad + \nabla_a\mathcal{E}^b\nabla_b\mathcal{E}^a + \nabla_b\mathcal{E}^a\,\nabla_a\mathcal{E}^b + \nabla_b\mathcal{E}^a\,\nabla^b\mathcal{E}_a\Big](\xi) \\ &= \int d^2\xi\sqrt{g}\Big\{\mathcal{E}_b(-\nabla^a\nabla_a)\mathcal{E}^b - \mathcal{E}^b(\nabla_a\nabla_b)\mathcal{E}^a \\ &\qquad - \mathcal{E}^a(\nabla_b\nabla_a)\mathcal{E}^b - \mathcal{E}^a(\nabla_b\nabla^b)\mathcal{E}_a\Big\} \\ &= -\int d^2\xi\sqrt{g}\Big\{2\mathcal{E}_b(-\nabla_a\nabla_a)\mathcal{E}^b - \mathcal{E}^b\,\overbrace{(\nabla_a\nabla_b + \nabla_a\nabla_b)\mathcal{E}^a}^{2\nabla_a\nabla_b=2\nabla_b\nabla_a+2[\nabla,\nabla^b]_-} \\ &= -2\int d^2\xi\sqrt{g}\,\mathcal{E}_a\Big[(\nabla^c\nabla_c)\delta_{ab} + \overbrace{[\nabla_b,\nabla_a]_-}^{(\nabla_a\mathcal{E}^a=0)}\Big]\mathcal{E}_b \end{aligned} \tag{2.44}$$

As an exercise to our diligent readers, the form of the above elliptic operator in the conformal gauge are given by (action on two-component objects)

$$\begin{aligned} \Delta_{(g=e^\varphi\delta_{ab})} &= e^{-\varphi(\xi)}\,\partial_{\bar{z}}\,\partial_z \\ \widehat{\mathcal{L}}_{(g=e^\varphi\delta_{ab})} &= e^{-2\varphi(\xi)}\,\partial_{\bar{z}}(e^{\varphi(\xi)}\,\partial_z) \end{aligned} \tag{2.45}$$

where we have introduced the complex plane notation and the above displayed operators now acting on complex functions (isomorphic to two-component objects = vector fields)

$$\frac{\partial}{\partial z} = \frac{\partial}{\partial \xi_1} - i\,\frac{\partial}{\partial \xi_2} \qquad (i = \sqrt{-1})$$
$$\frac{\partial}{\partial \overline{z}} = \frac{\partial}{\partial \xi_1} + i\,\frac{\partial}{\partial \xi_2} \tag{2.46}$$

In ref[3], we present the explicit evaluation of elliptic operators of the form (for j being a positive integer) below considered

$$\mathcal{L}_j = e^{-(j+1)\varphi(\xi)}\,\partial_{\overline{z}}(e^{j\varphi)\xi)}\,\partial_z).$$

The result is

$$\det{}_F(\mathcal{L}_j) = \exp\Big\{ -\frac{1+6j(j+1))}{12\pi}\int_\Omega d^2\xi\,\frac{1}{2}\,(\partial_a\varphi)^2(\xi) - \lim_{\varepsilon\to 0^{-1}}\frac{1}{2\pi\varepsilon}\int_\Omega d^2\xi\,e^{\varphi(\xi)}\Big\} \tag{2.47}$$

Let us point out the validity of the relationship below

$$\det{}_F{}^{-D/2}(\Delta_{g=e^\varphi\delta_{ab}}) = \det{}_F{}^{-D/4}\,(\mathcal{L}_{j=0})$$
$$\det{}^{+\frac{1}{2}}\,(\widehat{L}_{g=e^\varphi\delta_{ab}}) = \det{}_F{}^{+\frac{1}{2}}\,(\mathcal{L}_{j=1}) \tag{2.48}$$

As our final result of this covariant Polyakov path integral, we get the 2D-induced quantum gravity Liouville model

$$I = \int D^F\left[e^{\frac{\varphi}{2}(\xi)}\right]\exp\Big\{-\frac{(26-D)}{48\pi}\int_\Omega \frac{1}{2}\,(\partial_a\varphi)^2(\xi)\,d^2\xi\Big\} \times \Big\{e^{-\mu^2_{ren}}\int_\Omega e^{\varphi(\xi)}\,d^2\xi\Big\}. \tag{2.49}$$

Here $\mu_r^2 = \mu_0^2 + \lim\limits_{\varepsilon\to 0+}\dfrac{(2-D)}{8\pi\varepsilon}$.

Note that the quantum measure on the above displayed Liouville model is not the usual flat Feynman measure $\prod\limits_{\xi\in\Omega} d\varphi(\xi) = D^F[\varphi(\xi)]$ as initially wrongly supposed by A.M. Polyakov.

As a consequence one should re-write it in terms of the canonical Goldstone boson field $\overline{\rho}(\xi) = \left|\left(e^{\varphi(\xi)}\right)^{1/2}\right|$ which lead us to a kind of "non-compact" σ-model with a "renormalized" mass term

$$I = \int D^F[\overline{\rho}(\xi)]\exp\Big\{-\frac{(26-D)}{12\pi}\int_\Omega \frac{(\partial_a\overline{\rho})^2}{\overline{\rho}^2}\,d^2\xi - \mu_r^2\int_\Omega \overline{\rho}^2(\xi)\,d^2\xi\Big\} \tag{2.50}$$

and signaling the dynamical breaking of the conformal group ([1]).

Sometimes one can formally consider the variable change in the path-measure of eq. (2.49)

$$
\begin{aligned}
D^F\left[e^{\frac{\varphi}{2}(\xi)}\right] &= \overbrace{\det_F\left(e^{\frac{\varphi}{2}}\,\delta_{ab}\,\delta(\xi-\xi')\right) D^F[\varphi(\xi)]}^{\text{Functional Jacobian}} \\
&= \left(\lim_{\varepsilon\to 0^+} \exp\, Tr\left[\ell g\left(e^{\frac{\varphi}{2}}\,\delta_{ab}\right) e^{-\mathcal{E}\Delta}\, g = e^{\varphi}\right]\right) \times D^F[\varphi(\xi)] \\
&= \lim_{\varepsilon\to 0^+} \exp\left\{\int_\Omega d^2\xi\, e^{\varphi(\xi)}\left(\frac{\varphi(\xi)}{2}\right)\left[\frac{1}{h\pi\varepsilon}+\frac{1}{12\pi}R(\varphi)\right]\right\} \\
&\quad \times D^F[\varphi(\xi)] \\
&= \lim_{\varepsilon\to 0^+}\left\{\exp\left[\frac{1}{8\pi\varepsilon}\int_\Omega e^{\varphi(\xi)}\,\varphi(\xi)d^2\xi\right]\right. \\
&\quad \left.\times \exp\left[-\frac{1}{24\pi}\int_\Omega (\varphi(-\Delta)\varphi(\xi)))\xi)d^2\xi\right]\right\} \\
&\qquad \times D^F[\varphi(\xi)]
\end{aligned} \tag{2.51}
$$

which would lead to the same expression eq. (2.50) but with the "reduced" anomaly conformal coefficient 26-D to 25-D, if one could disregard the infinite piece on eq. (2.51) and if the "Fujikawa-like" evaluation of the functional jacobian could be done more invariant. At this point it appears that the above 2D-Liouville-Polyakov path integral only makes sense at the classical level which is formally equuivalent to evaluate all the observables in the $\varphi(\xi)$-theory at the limit of $D\to-\infty$ ([1]). One can see this after considering the re-scale

$$
\varphi \to \sqrt{\frac{48\pi}{26-D}}\,\varphi
$$

$$
D^F[e^{\frac{\varphi}{2}}] \to D^F\left[e^{\sqrt{\frac{48\pi}{26-D}}\,\varphi}\right] \overset{D\to-\infty}{\longrightarrow} D^F[e^0] \sim \text{classical.}
$$

However we should point out that all there questions on two-dimensional quantum gravity as a well-defined problem still are not well-understood since tis inception on 1981 ([1]).

Finally it is very important on applications to the Dual model theory for strong interactions (off-shell Scattering Amplitudes) to have a covariant regularized form for the formal Green function of the Beltrami-Laplace operator in the conformal gauge $g_{ab}(\xi) = e^{\varphi(\xi)}\,\delta_{ab}$. It is well know from Hadamard theory of parametrix singular solutions that one has the following covariant behavior for the Beltrami-Laplace operator for small distances

$$(\Delta^{-1}_{g=e^{\varphi}\delta_{ab}})(\xi,\xi')^{\varepsilon=(\xi-\xi')\to 0} \sim \frac{1}{2\pi\varepsilon} - \frac{1}{2\pi}\ell g(e^{\varphi(\xi)}|\xi-\xi'|)$$
$$\sim \frac{1}{2\pi}\left(\frac{1}{\varepsilon}\right) - \frac{1}{2\pi}\varphi(\xi) - \frac{1}{2\pi}\ell g|\xi-\xi'|. \tag{2.52}$$

This lead us to consider the regularized from of the Green function used on dual models based on the Polyakov's path integral

$$\left(\xi\left|\Delta^{-1}_{g=e^{\varphi}\,\delta_{ab}}\, e^{-\mathcal{E}\Delta_{g=e^{\varphi}\,\delta_{ab}}}\right|\xi'\right) = \begin{cases} -\frac{1}{2\pi}\ell g|\xi-\xi'| & \xi\neq\xi' \\ -\frac{1}{4\pi}\ell_g\left(\frac{1}{\varepsilon}\right) + \frac{1}{4\pi}\varphi(\xi) & \xi=\xi' \end{cases} \tag{2.53}$$

Note that we have used the following formulae to write explicitly the functional trace of the general two-dimensional (strongly) elliptic operators ([3])

$$A = -\left(g_{11}\frac{\partial^2}{\partial\xi_1^2} + g_{22}\frac{\partial^2}{\partial\xi_2^2}\right)\mathbf{1}_{2\times 2} - (A_1)\frac{\partial}{\partial x_1} - (A_2)\frac{\partial}{\partial x_2} - A_0 \tag{2.54-a}$$

$$\lim_{t\to 0^+} tr(e^{-tA})^{t\to 0^+} = \frac{1}{4\pi t}\left(\int d^2\xi\,(g_{11}g_{22})^{-\frac{1}{2}}(\xi)\right)$$
$$+\frac{1}{4\pi}\Big\{\int d^2\xi(g_{11}g_{22})^{-\frac{1}{2}}(-\frac{1}{6}R(g^{-1}))$$
$$-\frac{1}{2}\int d^2\xi(g_{11}g_{22})^{-\frac{1}{2}}\left(\frac{\partial}{\partial\xi^1}(g_{11}g_{22})^{-\frac{1}{2}}A_1\right)$$
$$+\frac{\partial}{\partial\xi^2}((g_{11}g_{22})^{-\frac{1}{2}}A_2) - \frac{1}{4}(A_1^2/g_{11}) - \frac{1}{4}(A_2^2/g_{22}) + A_0\Big\} + O(t) \tag{2.54-b}$$

2.4 Path-Integral quantization of the Nambu-Goto theory of random surfaces

In order to given a path integral meaning for the symbolic Feynman continuum sum over surfaces histories eq. (2.24-a), we start re-writting it in the following Polyakov's form, but in the presence of a constraint ([1])

$$G_{NG}(C) = \int d\overline{\mu}[x^{\mu}(\xi), g_{ab}(\xi)]\exp\left\{-\frac{1}{2\pi\alpha'}\int_{\Omega} d^2\xi(\sqrt{g}\,g^{ab}\,\partial_a X^{\mu}\,\partial_b\,X_{\mu})(\xi)\right\}$$
$$\times\,\delta^{(F)}_{cov}(g_{ab} - \partial_a X^{\mu}\partial_b X_{\mu}) \tag{2.55}$$

Note that our constraint is re-written as a covariant (diffeomorphism) invariant delta functional

$$\delta_{cov}^{(F)}(g_{ab} - \partial_a X^\mu \partial_b X_\mu) = \int D^{cov}[\lambda_{ab}] e^{i\{\int_\Omega d^2\xi(\sqrt{g}\lambda_{ab}(\partial^a X^\mu \partial^b X^\mu - g^{ab}))(\xi)\}} \tag{2.56}$$

where the covariant Feynman-Wiener measure on eq. (1.56) is the element of volume of the functional metric

$$ds^2 = ||\delta\lambda_{ab}||^2 = \int (\sqrt{g}\,\delta\,\lambda_{ab}(g^{aa'}\,g^{bb'})\delta\,\lambda_{a'b})(\xi)d^2\xi \tag{2.57}$$

After introducing again the complex-euclidean light-cone coordinates on the domain Ω

$$\begin{aligned} \partial_+ &= (\partial_{\xi_1} - i\,\partial_{\xi_2})/2 = \partial_z \\ \partial_- &= (\partial_{\xi_1} + i\,\partial_{\xi_2})/2 = \partial_{\overline{z}} \end{aligned} \tag{2.58}$$

and imposing the surface conformal gauge $g_{ab}(\xi) = \rho(\xi)\delta_{ab}$, which means that $g_{11} = g_{22}\,;\, g_{12} = g_{21} = 0$, or equivalently using light-cone coordinates again:

$$\begin{cases} \partial_+ X^\mu\,\partial_+ X_\mu = \partial_- X^\mu\,\partial_- X_\mu & (\Leftrightarrow \partial_{\xi_1} X^\mu \partial_{\xi_2} X_\mu = g_{12} = g_{21} = 0) \\ \sqrt{g} = \partial_+ X^\mu\,\partial_- X^\mu & (g_{11} = \partial_+ X^\mu\,\partial_- X_\mu = g_{22} \neq 0) \end{cases} \tag{2.59}$$

Now by taking into account the Jacobian relationship between the covariant functional measures and the associated pure Feynman-Wiener-Kac path measures, one obtains the following results:

A carefull discussion presented in ref[1]-chapter 12, lead us to the following result on the surface light-cone gauge eq. (2.59)

$$\begin{aligned} G_{NG}(C) &= \int D^F[\overline{\rho}(\xi)] D^F[X^\mu(\xi)] \\ &\times \exp\left\{-\frac{(26-D)}{12\pi}\int_\Omega d^2\xi\left[\frac{1}{2}\left(\frac{\partial_a\overline{\rho}}{\overline{\rho}}\right)^2\right] + \lim_{\varepsilon\to 0}\frac{(2-D)}{2\pi\varepsilon}\int_\Omega d^2\xi\,\overline{\rho}^2\right\} \\ &\times \exp\left\{-\frac{1}{2\pi\alpha'}\int_\Omega d^2\xi(\partial_a X^\mu)^2(\xi)\right\} \\ &\times \delta^{(F)}\left[(\overline{\rho})^2(\xi)\delta_{ab} - \overbrace{h_{ab}(X^\mu(\xi)}^{=\partial_+ X^\mu \partial_- X_\mu}\right] \end{aligned} \tag{2.60}$$

Note the usual non-covariant function "flat" constraint relating the surface vector position to the auxiliary metric field $g_{ab}(\xi)$ on the conformal gauge.

After realizing the immediate $\overline{\rho}(\xi)$ functional integral, one gets the final result, after a renormalization of the Regge slope constant ($\partial_+ = \partial_z, \partial_- = \partial_{\overline{z}}$) (*) [1]

$$G_N(C) = \int \left[\prod_{(\mu,\xi,\xi')} dX^\mu(z,\overline{z}) \sqrt[4]{h(X^\mu(z,z))} \right]$$
$$\times\, \delta^{(F)}(\partial_+ X^\mu \partial_+ X_\mu) \delta^{(F)}(\partial_- X^\mu \partial_- X^\mu) \partial X^\mu(\xi) e^\mu(\xi)$$
$$\exp\left\{ -\frac{1}{2\pi\alpha'_{ren}} \int_\Omega dz d\overline{z} (\partial_+ X^\mu)(\partial_- X_\mu)(z,\overline{z}) \right\}$$
$$\exp\left\{ -\frac{(26-D)}{48\pi} \int_\Omega dz\, d\overline{z} \left[\frac{\frac{1}{2}\partial_+(\partial_+ X^\mu \partial_- X_\mu)}{(\partial_+ X^\mu \partial_- X_\mu)} \times \frac{\partial_-(\partial_+ X^\mu \partial_- X_\mu)}{(\partial_+ X^\mu \partial_- X_\mu)} \right] (z,\overline{z}) \right\} \tag{2.61}$$

Now we realize that one most choose $D = 26$ or introduce intrinsic Majorance fermions living on those Nambu-Goto random surfaces in order to vanish the non-renormalizable Liouville surface term. The number of such Neutral Majorana 2D fields must be such that $26 = 4 + m$ on $D = 4$. Note that complex pairing such neutral Majorance fermion one gets eleven complex fermion Dirac fields living on the random surface.

It is worth (all the reader attention that in this framework of Nambu-Goto path integrals on surface light-cone gauge, the vertexs are given by object:

$$V = \left\{ \sqrt{h}\, h^{[a_1,a_2]} \ldots h^{a_{2m-1},a_{2n}} \right\} (X^\mu(z.\overline{z}))$$
$$\times \left\{ \partial_{a_1} X^\mu \ldots \partial_{a_{2n}} X^{\mu_{2\mu}} \cdot \overbrace{\mathcal{E}_{\mu_1 \ldots \mu_{2m}}}^{\text{spin polarization tensor}} \cdot e^{iK_\mu X^\mu} \right\} (z,\overline{z})$$

which are free from tachions, etc...

A complete study of the spectrum of such Nambu-Goto strings with those tecnions-free "geometrical" vertexs is still missing on literature and left to our diligent readers.

Similar analysis can be straightforwardly implemented for the extrinsic surface path-integral weight functional eq. (2.17)–eq. (2.18).

[1](*) It is worth to remark that at a Feynman diagrammatic perturbative level, the measure "tad-pole" factor in eq. (2.61), $\exp\left\{ \frac{-\delta^{(2)}(0)}{4} \int_\Omega \ell g(h(X^\mu(\overline{z}))) \times dz d\overline{z} \right\}$, can be assigned to $\exp\{0\} = 1$ in the dimensional regularization scheme.

2.5 References

[1] Luiz C.L. Botelho - Methods of Bosonic and Fermionic Path Integrals Representation - Continuous Random Geometry in Quantum Field Theory - Nova Science - (2007)
[2] Luiz C.L. Botelho - Lecture Notes in Applied Differential Equations of Mathematical Physics - World Scientific - 2008.
[3] J.T. Oden and J.N. Reddy - An Introduction to the Mathematical Theory of finite elements - John Willey & Sons - 1976.

2.6 Appendix A: 2D Abelian Dirac Determinant On the Formal Evaluation of the Euclidean Dirac Functional Determinant on Two-dimensions

Let us start by this appendix by considering the euclidean Dirac partition functional on a tw-dimensional space-time

$$Z[A_\mu] =: \frac{1}{Z[A_\mu = 0]} \times \left\{ \int D\psi\, D\overline{\psi}\, e^{-\int (\overline{\psi} \not{D}(A)\psi)(\xi) d^2\xi} \right\}$$
$$=: \det(\not{D}(A)) = \det{}^{1/2}((\not{D}(A))^2) \tag{A-1}$$

where the (non-self adjoint) euclidean Dirac operator is explicitly given by (including the massive case!)

$$\not{D}(A) = \gamma_\mu(\partial_\mu + g\, A_\mu) + im. \tag{A-2}$$

The two-dimensional euclidean matrixes on the Dirac operator eq. (A-2) satisfy the relationship below

$$\{\gamma_\mu, \gamma_\nu\} = \gamma_\mu\, \gamma_\nu + \gamma_\nu\, \gamma_\mu = 2\delta_{\mu\nu}$$
$$\gamma_\mu\, \gamma_5 = i\, \mathcal{E}_{\mu\nu}\, \gamma_\nu\,, \; \gamma^5 = i\, \gamma_0\, \gamma_1$$
$$\mathcal{E}_{01} = -\mathcal{E}_{10} = 1 \tag{A-3}$$

Let us note the more suitable re-definition of the euclidean Dirac operator

$$\widetilde{\not{D}}(A) = i\not{D}(A) = \overbrace{e^{ig\gamma_5 \not{D}}(i\,\gamma_\mu \partial_\mu) e^{ig\gamma_5 \not{D}}}^{=i\,\gamma_\mu \partial_\mu + ig\,\partial_\mu A_\mu} \tag{A-4}$$

where the back-ground abelian gauge field configurations are choosen on the landau gauge $\partial_\mu\, A_\mu = 0$

$$A_\mu = \mathcal{E}_{\mu\nu}\, \partial_\nu\, \varphi \tag{A-5}$$

Let us note that (with a undetermined infinite phase!)

$$\frac{\det_F(\widetilde{\not{D}}(A))}{\det_F(\widetilde{\not{D}}(0))} = \frac{\det_F(\widetilde{\not{D}}(A))}{\det_F(\widetilde{\not{D}}(0))} \tag{A-6a}$$

$$\left|\frac{\det_F(\widetilde{\not{D}}(A))}{\det_F(\widetilde{\not{D}}(0))}\right|^2 = \left|\frac{\det_F(\widetilde{\not{D}}(A))}{\det_F(\widetilde{\not{D}}(0))}\right|^2 \tag{A-6b}$$

Since we have the result:

$$(\widetilde{\not{D}}(A))^2 = -\,\partial^2 - 2g\,A_\mu(\xi)\partial_\mu - g^2\,A_\mu^2(\xi) - \frac{g}{4}\,\overbrace{\sigma^{\mu\nu}}^{[\gamma^\mu,\gamma^\nu]}\cdot F_{\mu\nu}(A(\xi)) - g(\partial_\mu\,A_\mu)(\xi) \tag{A-7}$$

thus

$$\lim_{t\to 0^+}\left\langle\left|\left\{\exp(-t(\widetilde{\not{D}}(A))^2)\right\}\right|\xi\right\rangle \sim \frac{1}{4\pi t}\,\mathbf{1}_{2\times 2} + \frac{1}{4\pi}\left[g^2A_\mu^2 + \frac{g}{4}\,\sigma^{\mu\nu}F_{\mu\nu} + g\,\partial_\mu A_\mu - \frac{1}{4}(2gA_\mu)^2 - \frac{1}{2}\,\partial_\mu(2gA_\mu)\right], \tag{A-8}$$

one can straightforwardly apply the Romanov-Schwartz theorem ([1]) to write the following (formal) differential equation for the functional determinant under analysis

$$2\,\frac{d}{d\sigma}\,\ell_g\,\det(\not{D}(A^\sigma)) = \lim_{\varepsilon\to 0^+} 4\left\{\int d^2\xi\,\mathrm{Tr}_{\mathrm{Dirac}}(ig\gamma^5\varphi)\left[\frac{1}{4\pi\varepsilon} + \frac{g}{16\pi}\sigma\,\overbrace{\sigma^{\mu\nu}}^{=[\gamma^\mu,\gamma^\nu]}\,F_{\mu\nu}(A)\right]\right\} \tag{A-9}$$

Here

$$A^{(\sigma)} = \sigma\,\mathcal{E}_{\mu\nu}\,\partial_\nu\,\varphi \tag{A-10}$$

Since $\mathrm{Tr}_{\mathrm{Dirac}}(\gamma_5) = 0$, one has the immediate result

$$\left(\frac{\det\not{D}(A)}{\det\,\not{\partial}}\right) = e^{-\frac{g^2}{2\pi}\int(\partial\phi)^2\,d^2\xi} = e^{-\frac{g^2}{2\pi}\int A_\mu^2(\xi)d^2\xi} \times \delta^{(F)}(\partial_\mu\,A_\mu) \tag{A-11}$$

In the general non gauge fixed case, one has the result:

$$\frac{\det(\not{D}(A))}{\det(\not{\partial})} = \int D^F[\omega(x)]e^{-\frac{g^2}{2\pi}\int(A_\mu+\partial_\mu\omega)^2d^2\xi} = e^{-\frac{g^2}{2\pi}\int d^2\xi\left[A_\mu\left(\delta_{\mu\nu}-\frac{\partial_\mu\partial_\nu}{(-\partial^2)}\right)A_\nu\right](\xi)} \tag{A-12}$$

Let us show in detail the above calculation

$$\int_0^1 (\ell_g \det(\not{D}(A^{(\sigma)}))^2) d\sigma$$

$$= \lim_{\varepsilon\to 0^+} 4\left\{\int d^2\xi \operatorname{tr}\left(ig\gamma^5\varphi\left[\frac{1}{4\pi\varepsilon} + \frac{g}{16}\sigma^{\mu,\nu} F_{\mu\nu}(A)\right]\right)\right\} \times \left(\int_0^1 d\sigma\,\sigma\right).$$

So

$$\begin{aligned}
\ell_g\left(\frac{\det(\not{D}(A))}{\det(\not{\partial})}\right) &= \frac{g^2}{16\pi}\left\{\int d^2\xi \mathrm{Tr}_{\mathrm{Dirac}}(i\gamma^5(-2i\,\mathcal{E}_{\mu\nu})\gamma_5\phi \times F_{\mu\nu}(A)\right\} \\
&= \frac{g^2}{4\pi}\int d^2\xi\, \phi\, \mathcal{E}_{\mu\nu}\, F_{\mu\nu}(A) \\
&= \frac{g^2}{2\pi}\int d^2\xi\, \phi\, \partial^2\phi = -\frac{g^2}{2\pi}\int d^2\xi(\partial\phi)^2 = -\frac{g^2}{2\pi}\int d^2\xi(G_\mu)^2
\end{aligned} \tag{A-13}$$

It is important to remark that if one had used the following self-adjoint Dirac euclidean operator

$$\widehat{\not{D}}(A) = e^{e\gamma_5\phi}(i\gamma_\mu\partial_\mu)e^{g\gamma_5\phi} \tag{A-14}$$

one would obtain a "Tachion" (imaginary mass) term for the dynamically generated mass term for the abelian gauge field

$$\frac{\det_F \widehat{\not{D}}(A)}{\det\,\not{\partial}} = \exp\left\{+\frac{g^2}{2\pi}\int d^2\xi(G_\mu)^2\right\} \tag{A-15}$$

At this point it is worth recall the Seeley asymptotic expansion

$$\lim_{t\to 0^+} \langle\xi|\exp(-tA)|\xi\rangle$$

$$= \frac{1}{4\pi t}\mathbf{1}_{2\times 2} + \frac{1}{4\pi}\left[V_0 - \frac{V_\mu^2}{4} - \frac{\partial_\mu V_\mu}{2}\right] \tag{A-16}$$

for a formal second order operator of the form on R^2

$$A = -\partial^2 - V_\mu\partial_\mu - V_0 \tag{A-17}$$

2.7 Appendix B: On Atyah-Singer Index Theorem in the Framework of Feynman Pseudo-Classical Path Integrals - Author's Original Remarks

One of the most celebrated (pure) mathematical theorem on modern geometry and topology of compact orientable manifolds on R^N is the so called index theorem ([1]). By the other side, it is widely known that the techniques to obtain such results are extremelly intricated and based on the most difficult aspects of the theory of elliptic operators on closed compact orientable N-dimensional manifolds (see [2]-chapter 11). Our aim in this section is to somewhat simplify a very important index theorem on the Laplacian operator acting on forms ([3]).

As in the last reference [3], one can argue that the index of the Laplacian operator Δ_f acting on differential forms on a N-dimensional Riemannian compact orientable manifold $\mathcal{M}$ is given by the object called super-trace which by its turn can be wirtten as a Grasnmanian (supersymmetric) euclidean path integral ([1]):

$$\text{index}(\Delta_f) =: \lim_{\hbar\to 0}\left\{\int\left(\overbrace{\prod_{\substack{0\le s\le 1\\ 1\le n\le N}} \det{}^{\frac{1}{2}}[G_{\mu\nu}(X^\mu(s))]dX^\mu(s)}^{\equiv D^F_{\text{cov}}[X^\mu(s)]}\right)\right\}$$

$$X^\mu(0) = X^\mu(1) = x^\mu$$

$$\times\int\left(\overbrace{\prod_{\substack{0\le s\le 1\\ \alpha=1,2\\ 1\le i\le 2N}} \det{}^{\frac{1}{2}}[\sqrt{G}(X^\mu(s))]d\xi^i_{(\alpha)}(s)d\xi^{*i}_{(\alpha)}(\alpha)}^{\equiv D^F_{\text{cov}}(\xi^i_{(\alpha)}(\xi)]}\right)$$

$$\exp\left\{-\frac{1}{\hbar}\int_0^1 ds\left[\frac{1}{2}G_{\mu\nu}(X^\alpha(s))\frac{dX^\mu}{ds}(s)\frac{dX^\mu}{ds}(s)\right.\right.$$
$$+\frac{i}{2}G_{\mu\nu}(X^\alpha(s))\left(\xi^i_{(\alpha}(s)\frac{d}{ds}\xi^i_{(\alpha)}\right)$$
$$\left.\left.+\frac{1}{8}R_{ijk\ell}(X^\alpha(s))(\xi^i_{(1)}(s)\xi_{(2)}(s)^k)(\xi^j_{(1)}(s)\xi^\ell_{(2)}(s))\right]\right\} \tag{B-1}$$

An important remark to be done now originally due to ourselves - [4]) is about the "bosonic" behavior of the Grassmanian composite products $\xi^i_{(1)}(s)\,\xi^j_{(2)} = \sigma^{ij}(s)$ as can seen easily verified from the computation below ($[a,b]_- = ab - ba$; $[a,b]_+ = ab + ba$) $-[AB,CD]_- = A[B,C]_+D - AC[B,D]_+ + [A.C]_+DB - C[A,D]_+B$

$$\begin{aligned}\left[\xi^i_{(1)}\,\xi^j_{(2)},\xi^p_{(1)}\,\xi^q_{(2)}\right]_- &= \xi^i_{(1)}\left[\xi^p_{(1)}\,\xi^q_{(2)},\xi^j_{(2)}\right]_+ - \left[\xi^p_{(1)}\,\xi^q_{(2)},\xi^i_{(1)}\right]_+\xi^j_{(2)}\\ &= \left\{\xi^i_{(j)}\left(\xi^p_{(1)}\,\xi^q_{(2)}\,\xi^j_{(2)}\right) - \xi^p_{(1)}\,\xi^q_{(2)}\,\xi^i_{(1)}\,\xi^j_{(2)}\right\}\\ &\quad+\left\{\xi^i_{(1)}\,\xi^j_{(2)}\,\xi^p_{(1)}\,\xi^q_{(2)} - \xi^i_{(1)}\,\xi^p_{(1)}\,\xi^q_{(2)}\,\xi^j_{(2)}\right\}\\ &= 0+0=0\end{aligned} \tag{B-2}$$

On the basis of the above made remark eq. (B-2) one can re-write the Grasmanian path integral eq. (B-1) in the terms of "composite bosonic operators" ([4])

$$\begin{aligned}\text{index}(\Delta_f) = \lim_{t\to 0} \int_{X^\mu(0)=X^\mu(1)=\overline{X}^\mu} D^F_{\text{cov}}[X^\mu(s)] \int_{\xi^i_\alpha(0)=-\xi^i_\alpha(1)} D^F[\xi^i_\alpha(s)] \\ \times \int_{\sigma_{ik}(0)=\sigma_{ik}(1)} D^F_{\text{cov}} \overbrace{[\sigma_{ik}(s)]}^{\text{Bosonic Variable}} \int_{\lambda_{ik}(0)=\lambda_{ik}(1)} D^F \overbrace{[\lambda_{ik}(s)]}^{\text{Bosonic Variable}} \\ \times \exp\left\{\frac{i}{h}\int_0 ds\,\lambda^{ik}(\xi^i_{(1)}(s)\,\xi^k_{(2)}(s) - \sigma_{ik}(s))\right\} \\ \times \exp\left\{-\frac{1}{4h}\int_0^1 \left(\frac{1}{2}R_{ijk\ell}(X^\alpha(s))\sigma^{ik}_{(s)}\,\sigma^{j\ell}_{(s)}\right)\right\}\end{aligned} \tag{B-3}$$

After integrating the Grasmanian variables $\{\xi^i_{(1)}(s), \xi^i_{(2)}\}_{1\le i\le N}$, one gets the below written functional determinant as outcome (prove it with the definition $\theta(0) = 0$, here $\theta(s)$ denotes the heaviside step function-distribution)

$$\det_{\text{cov}}\left[G_{ij}(X^\alpha(s))\frac{d}{ds}\delta_{ij} + \lambda_{ik}(s)\right] = 1 \tag{B-4}$$

Now the σ_{ik} functional integral is Gaussian and yielding the result

$$\exp\left\{-\frac{1}{2\hbar}\int_0^1 ds\int_0^1 ds'(\lambda_{ik}(s))\left[\frac{1}{2}R_{pqrs}(X^\alpha_{(s)})\right]^{-1}_{(ik).(j\ell)}\delta(s-s')(\lambda_{j\ell}(s))\right\} \tag{B-5}$$

At the classical limit $\hbar\to 0$, the closed periodic quantum trajectories $X^\mu(0) = X^\mu(s) = \overline{x}^\mu$, all reduces to theirs initial common point $\overline{x}^\mu$. As a consequence one gets that at $\hbar \to 0$.

$$X^\mu(s) \to \overline{x}^{-\mu} \tag{B-6}$$

$$D^F_{\text{cov}}(X^\mu(s)) \to \int_{\mathcal{M}} d^N(\overline{x})\cdot \det^{\frac{1}{2}}[G_{\mu\nu}(\overline{x}^{-\alpha}] \tag{B-7}$$

$$\begin{aligned}\int D^F_{\text{cov}}(\lambda_{pq}(s)]\exp\Big\{-\frac{1}{\hbar}\int_0^1 ds\int_0^1 ds'(\lambda\underbrace{_{ik}}_{A}(s)) \\ [R_{(A,B)}(X^x(s)]^{-1}(\lambda\underbrace{_{j\ell}}_{B}(s'))\Big\} \\ \to \det^{\frac{1}{2}}[R_{AB}(\overline{x}^{-\mu})] =: P_f(R_{AB}(x^{-\mu}))\end{aligned} \tag{B-8}$$

By group together eqs. (B-6)–eq. (B-8), one gets the index of the Laplacean operator acting on differential forms defined on a compact orientable N-dimensional is universally proportional to the famous Chern index of the endowed riemannian structure of M. Namelly (see ref [3] for topological details). Plus analogous induced term on the Manifold Boundary if it is non empty (an open manifold).

$$\text{index}(\Delta_f) = \xi_{\text{Chern}}(\mathcal{M}) = \overline{e}\left\{\int_{\mathcal{M}} d^N(\overline{x})\det^{\frac{1}{2}}[G_{\mu\nu}(\overline{x}^\mu)] \times \overbrace{P_f(R_{AB}(\overline{x}^\mu)}^{\text{Pfaffinian form of the curvature two-form}}\right\} \quad \text{(B-9)}$$

The universal constant is adjusted by defining as unity the Chern induces or the Euler-Poincar characteristic) of spheres S^{N-1} with the usual R^N embedding metric (exercise to our differential geometric oriented readers!). Now it is well known that $\chi_{\text{Chern}}(\mathcal{M})$ are integers numbers.

In two dimensions, eq. (B-9) reduces to the famous Gauss theorem for Riemann surfaces and its conections to the Euler topological surface genera

$$\text{index}(\Delta_f) = \chi_{\text{Chern}}(\Sigma) = \int_{\Sigma} (\sqrt{g}R(g))(\overline{x}^\mu)d^2\overline{x} = 2\pi(2-2g), \quad \text{(B-10)}$$

where g is the number of "holes" of Σ.

2.8 References

[1] M. Nakahara - "Geometry, Topology and Physics" - Graduate Student Series in Physics - Institute of Physics, 1990, UK.

[2] Luiz C.L. Botelho - Lecture Notes in Applied Differential Equations of Mathematical Physics - World Scientific - 2008 (Singapure).

[3] P. Cartier and Ccile De Witt-Morette - Functional Integration - Action and Symmetries - Cambridge Monogrphs on Mathematical Physics - 2006, UK.

[4] Luiz C.L. Botelho - String Theory in Embeddings Manifolds - IJTP, vol. 49, Issue 8, 1886-1893, DOI = 10.1007/s10773-010-0371-9, (2010).

2.9 Appendix C: Path integral bosonisation for the Thirring model in the presence of vortices

Recently the bosonisation of two-dimensional fermionic models in the framework of path integrals has been shown to be a powerful non-perturbative technique for exactly solving these models in the absence of fermion zero modes [1,2].

It is the purpose of this Appendix to implement the path integral bosonisation framework for the case of non-trivial fermion zero modes by bosonisation of the Abelian Thirring model in the presence of vortex field configurations.

Let us start our analysis by considering the Abelian Thirring model interacting with a vortex field with topological charge n. Its Euclidean Lagrangian is given by

$$\mathcal{L}_0(\psi,\bar{\psi},A_\mu^{(n)}) = \bar{\psi}i\gamma_\mu\partial)\mu\psi + \frac{1}{2}g^2(\bar{\psi}\gamma_\mu\psi)^2 + e\bar{\psi}\gamma_\mu A_\mu^{(n)}\Psi \tag{C-1}$$

where (g,e) are positive model coupling constants. The Euclidean Hermitian γ_μ matrices we are using satisfy the relations

$$\{\gamma_\mu,\gamma_\nu\} = 2\delta_{\mu\nu} \quad \gamma_\mu\gamma_5 = e\varepsilon_{\mu\nu}\gamma_\nu \quad \gamma_5 = i\gamma_0\gamma_1 \quad \varepsilon_{01} = -\varepsilon_{10} = 1 \tag{C-2}$$

and the vortex field $A_\mu^{(n)}(x)$ with topological charge n (Chern number) and length' R is

$$A_\mu^{(n)}(x) = n\frac{1}{R^2+x^2}\varepsilon_{\lambda\mu}x_\lambda. \tag{C-3}$$

The generating functional associated with the theory (1) is, thus, given by

$$Z[\eta,\bar{\eta},A_\mu^{(n)}] = \int D[\psi]D[\bar{\psi}]\exp\left(-\int d^2x(\mathcal{L}_0(\psi,\bar{\psi},A_\mu^{(n)}) + \bar{\eta}\psi + \bar{\psi}\eta)(x)\right). \tag{C-4}$$

In order to implement the path integral bosonisation gauge invariant technique [1,2], we rewrite the fermion interaction term in the Hubbard-Stratonovitch form:

$$\begin{aligned} Z[\eta,\bar{\eta},A_\mu^{(n)}] = &\int D[\psi]D[\bar{\psi}]\exp\left(-\frac{1}{2}\int d^2xB_\mu^2(x)\right) \\ &\times\exp\left(-\int d^2x[\bar{\psi}i\gamma_\mu(\partial_\mu + B_\mu - ieA_\mu^{(n)})\psi](x)\right) \\ &\times\exp\left(-\int d^2x(\bar{\psi}\eta + \bar{\eta}\psi)(x)\right). \end{aligned} \tag{C-5}$$

Let us now proceed as in [1,2] by making the (partial) decoupling field change in (C-5)

$$\psi(x) = [\exp(ig\gamma_s u)x](x) \quad \bar{\psi}(x) = [\chi \exp(ig\gamma_s u)](x)$$
$$B_\mu(x) = (g\varepsilon_{\mu\nu}\partial_\nu u)(x). \tag{C-6}$$

As has been shown by Fujikawa [3], the transformations of (C-6) are not free of cost in the fermionic sector since the functional measure $D[\psi]D[\bar{\psi}]$ is defined in terms of the normalised eigenvectors of the Dirac operator $\not{D}(B_\mu, A_\mu^{(n)}$ in the presence of the external (auxiliary) Abelian field B_μ and of the vortex topological field $A_\mu^{(n)}$.

At this point we note that after the chiral change takes place, the new quantum fermion vacuum is defined by the fermionic theory in the presence of the topological vortex; i.e. $D[\chi]D[\tilde{\chi}]$ is now defined in terms of the eigenvectors of the Dirac operator $\not{D}(A_\mu^{(n)})$ which in turn has precisely n fermionic zero modes with definite chiriality [4]. Their explicit expressions are

$$\psi_{(0),l}(x) = \left(\frac{1}{2\pi R}\right)^{1/2}\left(\frac{x_1 + ix_2}{R}\right)^{l-1}\begin{pmatrix} h_-(x, A_\mu^{(n)} \\ 0 \end{pmatrix} \tag{C-7a}$$

$$\gamma_5\psi_{(0),l}(x) = \psi_{(0),l}(x) \tag{C-7b}$$

$$h_\pm(x, A) = \exp\left(ie\int d^2z\Delta_{m=0}(x-z)(\partial_\mu A_\mu^{(n)} \pm i\varepsilon_{\mu\nu}\partial_\mu A_\nu^{(n)})(z)\right) \tag{C-7c}$$

with $\Delta_{m=0}(z)$ the (infrared regularised) massless Klein-Gordon propagator. The associate Jacobians are given by [1,2]

$$D[\psi]D[\bar{\psi}] = D[\chi]D[\bar{\chi}]\,\frac{\det \not{D}(B_\mu, A_\mu^{(n)})}{\det \not{D}(B_\mu = 0, A_\mu^{(n)})}. \tag{C-8a}$$

$$D[B_\mu] = D[u] \times \det[-\partial^2]. \tag{C-8b}$$

So, we face the problem of the evaluation of the ratio of the two Dirac determinants with zero modes.

By following the procedure of [1,2], we first introduce a one-parameter continuous family of Dirac operators itnerpolating the pure vortex Dirac operator and $\not{D}(A_\mu^{(n)}, B_\mu) = i\gamma_\mu((\partial_\mu + gB_\mu + ieA_\mu^{(n)}$ defined by the expression

$$\not{D}^{(\sigma)}(B_\mu, A_\mu^{(n)}) = \exp(ig\gamma_5\sigma u)\not{D}(A_\mu^{(n)})\exp(ig\gamma_5\sigma u) \quad (0 \leq \sigma \leq 1). \tag{C-9}$$

By using a proper-time prescription to define the functional determinant of $\not{D}^{(\sigma)}(B_\mu, A_\mu^{(n)})$ (after making the analytic exrension $g = -i\bar{g}$ [2]). We have the result $\log\det \not{D}^{(\sigma)} = \frac{1}{2}\log\det(\not{D}^{(\sigma)^2})|_{g=ig}$

$$= -\lim_{\varepsilon\to 0^+}\int_\varepsilon^{1/\varepsilon}\frac{d\zeta}{\zeta}Tr[\exp(-\zeta\not{D}^{(\sigma)^2})(\mathbf{1}-\mathbb{P}^{(\sigma)})] \tag{C-10}$$

with

$$\mathbb{P}^{(\sigma)}|\zeta\rangle = \sum_{l=1}^{n}\langle\zeta|\beta^\sigma_{(0),l}\rangle\beta^\sigma_{(0),l}$$

denoting the projection over the zero modes of the Dirac operator $\not{D}^{(\sigma)}(B_\mu, A_\mu^{(n)})$. Let us remark that by the Atyah-Singer theorem, this operator still has n zero modes $\beta^\sigma_{(0),l}(x)$ which are related to those of (C-7a) by an analytically continued chiral rotation (C-6):

$$\beta^\sigma_{(0),l}(x) = \exp(-\bar{g}\gamma_5\sigma u)\psi_{(0),l}(x). \tag{C-11}$$

The functional determinant (C-10) satisfies the following differential equation:

$$\begin{aligned}&\frac{d}{d\sigma}\log\det\not{D}^{(\sigma)}(B_\mu, A_\mu^{(n)})\\ &\quad= -2\lim_{\varepsilon\to 0^+}Tr[\bar{g}\gamma_5 u\exp(-\zeta\not{D}^{(\sigma)^2})(\mathbf{1}-\mathbb{P}^{(\sigma)})]\Big|_\varepsilon^{1/\varepsilon}\\ &\qquad+\lim_{\varepsilon\to 0^+}\int_\varepsilon^{1/\varepsilon}\frac{d\zeta}{\zeta}Tr\left[\exp(-\zeta\not{D}^{(\sigma)^2})\frac{d}{d\sigma}\mathbb{P}^{(\sigma)}\right].\end{aligned} \tag{C-12}$$

The second term in the right-hand side of (C-12) is easily evaluated by noting that

$$\frac{d}{d\sigma}\mathbb{P}^{(\sigma)}|\zeta\rangle = \sum_{l=1}^{m}\langle\zeta|-\bar{g}\gamma_5 u\beta^\sigma_{(0),l} + \sum_{l=1}^{n}\langle\zeta|\beta^\sigma_{(0),l}\rangle - \bar{g}\gamma_5 u\beta^\sigma_{(0),l} \tag{C-13}$$

which yields the zero-mode contribution for the determinant (12):

$$\begin{aligned}&\lim_{\varepsilon\to 0^+}\int_\varepsilon^{1/\varepsilon}\frac{d\zeta}{\zeta}Tr\left(\exp(-\zeta\not{D}^{(\sigma)^2})\frac{d}{d\sigma}\mathbb{P}^{(\sigma)}\right)\\ &\quad= \lim_{\varepsilon\to 0^+}\left[4log(\varepsilon)\bar{g}\sum_{l=1}^{n}\langle\beta^\sigma_{(0),l}u\beta^\sigma_{(0),l}\rangle\right].\end{aligned} \tag{C-14}$$

Since $\not{D}^{(\sigma)^2}$ is a self-adjoint invertible operator in the manifold orthogonal to its zero modes, wed can use the usual Seeley-De Witt technique to evaluate the first term in (C-12), which produces the usual result

$$\begin{aligned}&Tr(\bar{g}\gamma_5 u\exp(-\zeta\not{D}^{(\sigma)^2})(\mathbf{1}-\mathbb{P}^{(\sigma)}))\Big|_\varepsilon^{1/\varepsilon}\\ &\quad= \sigma\frac{1}{\pi}Tr\left((-i\bar{g})u(-i\bar{g})\partial^2 u + \frac{\varepsilon_{\mu\nu}}{2}F|_{\mu\nu}(A_\mu^{(n)})\right).\end{aligned} \tag{C-15}$$

By combining (C-15) with (C-14) and coming back to the real coupling constant g, we obtain the final expression for the Jacobian (8a) after integrating the differential equation (C-12)

$$\begin{aligned}
&\log\det \not{D}(B_\mu, A_\mu^{(n)}) - \log\det \not{D}(B_\mu = 0, A_\mu^{(n)}) \\
&\quad = \frac{g^2}{\pi}\int d^2x \frac{1}{2}(\partial u)^2 + \frac{g}{\pi}\int d^2x \varepsilon_{\mu\nu} F_{\mu\nu}(A_\mu^{(n)})u \\
&\quad - igC(\varepsilon)\sum_{l=1}^{n}\int d^2x \psi^+_{(0),l}(x)u(x)\psi_{(0),l}(x) \qquad \text{(C-16)}
\end{aligned}$$

where we have used the unitarity of the matrix $\exp\{i\bar{g}\gamma_5 u\}$ to evaluate (C-16) and $C(\varepsilon)$ is the usual infrared divergence contribution for the zero-mode term, which can be made finite by a multiplicative renormalisation of the Thirring coupling constant g.

The generating functional thus takes the simple form

$$\begin{aligned}
&Z[\eta, \bar{\eta}, A_\mu^{(n)}] \\
&\quad = \int \widehat{D[u]} \exp\Big\{ -\frac{1}{2}\int d^2x \Big[\Big(1 - \frac{g^2}{\pi}\Big)(\partial u)^2\Big](x) \\
&\quad - \frac{g}{\pi}\int d^2x (\varepsilon_{\mu\nu} F_{\mu\nu}(A_\mu^{(n)})u)(x) \\
&\quad \times \exp\Big(- ig\int d^2x \psi^+_{(0),l}(x)u(x)\psi_{(0),l}(x)\Big)\Big\} Z_{(0)}[\eta, \bar{\eta}, A_\mu^{(n)}, u]
\end{aligned} \qquad \text{(C-17)}$$

with $Z_{(0)}[\eta, \bar{\eta}, A_\mu^{(n)}, u]$ being the generating functional for the fermions in the pure vortex field configuration

$$\begin{aligned}
Z_{(0)}[\eta, \bar{\eta}, A_\mu^{(n)}, u] &= \int D[\chi]D[\bar{\chi}] \\
&\times \exp\Big(- \int d^2x (\bar{\chi}\not{D}(A_\mu^{(n)})\chi + \bar{\chi}\exp ig\gamma_5 u\eta + \bar{\eta}\exp ig\gamma_5 u\chi)(x)\Big).
\end{aligned} \qquad \text{(C-18)}$$

Let us exemplify our approach by calculating the two-point fermion correlation function for an external vortex field with topological charge 1. By functional differentiation of $Z_{(0)}[\eta, \bar{\eta}, A_\mu^{(n)}, u]$ twice, we get

$$\begin{aligned}
&(\bar{\chi}\exp ig\gamma_5 u)_\alpha(x)(\exp ig\gamma_5 u\chi)_\beta(y) \\
&\quad = [\exp(ig\gamma_5 u(x))S^{(1)}(x-y)\exp(ig\gamma_5 u(y))]_{\alpha\beta} \times \det(\not{D}(A_\mu^{(n)}))
\end{aligned} \qquad \text{(C-19)}$$

where $S^{(1)}(x-y)$ is the Euclidean Green function of the Dirac operator $\not{D}(A_\mu^{(n)})$ for $n=1$ which is given explicitly by [4]

$$S^{(1)}(x-y) = -\frac{1}{2\pi}\begin{bmatrix} 0 & \left(\frac{1+y^2}{1+x^2}\right)^{1/2}\frac{x_1-ix_2-(y_1-iy_2)}{(x-y)^2}+\frac{(y_1-iy_2)}{(1+x^2)^{1/2}(1+y^2)^{1/2}} \\ -\left(\frac{1+x^2}{1+y^2}\right)^{1/2}\frac{x_1+ix_2-(y_1+iy_2)}{(x-y)^2}+\frac{x_1+ix_2}{(1+x^2)^{1/2}(1+y^2)^{1/2}} & 0 \end{bmatrix}. \tag{C-20}$$

By evaluation of the u average of (C-19) we finally have the complete expression for $\langle\bar{\psi}(x)\psi|y)\rangle$:

$$\langle\bar{\psi}(x)\psi(y)\rangle = \Big[\exp\Big(\frac{1}{2}\frac{1}{1-g^2/\pi}\int d^2z d^2\bar{z} J_\mu(z;[x,y]) \times \Delta_{m=0}(z-\bar{z})J_\mu(\bar{z};[x,y])\Big)S^{(1)}(x,y)\Big] \tag{C-21}$$

where

$$J_\mu(z;[x,y]) = \Big(\frac{g}{\pi}\varepsilon_{\mu\nu}F_{\mu\nu}(A_\mu^{(1)}(z)) - ig\psi^+_{(0),0}(z)\psi_{(0),0}(z) + (ig\gamma_5^{(x)}\delta(z-x)+ig\gamma_5^{(y)}\delta(z-y))\Big) \tag{C-22}$$

with

$$\psi_{(0),0}(z) = \frac{1}{\sqrt{2\pi}}\left(\frac{1}{\sqrt{1+z^2}},0\right)$$

begin the only zero mode of $\not{D}(A_\mu^{(1)})$.

2.10 References

[1] Botelho L.C.L. (1985) *Phys. Rev.* D **31** 1503; 1986 *Phys. Rev.* D **33** 1195, **34** 3250.
[2] Botelho L.C.L. (1989) *Phys. Rev.* D **39** 3051.
[3] Fujikawa K. (1981) *Phys. Rev.* D **21** 2848.
[4] Rothe K.D. and Schroer B. (1979) *Phys. Rev.* D **20** 3203.

2.11 Appendix D:

Several attempts have been made to find a complete solution of two-dimensional quantum chromodynamics with massless fermions ([1], [2], [3]), shortly denoted by $(QCD)_2$. All these approaches were based on integrating out the fermion fields of the model and considering an effective gluonic theory where some interesting phenomena can be easily seen ([2], [3]).

In this appendix we intend to make some clarifying remarks on this effective action for massless $(QCD)_2$.

We start our analysis by considering the generating functional for the model in a Euclidean space-time R^2 with local gauge group $SU(2)$ (the generalization for the case $SU(N)$ is straightforward)

$$Z[J_\mu, \eta, \bar{\eta}] = \int D[G_\mu] e^{-\frac{1}{4g^2}\int d^2x Tr^{(c)}(F^2_{\mu\nu})(x)} e^{\int d^2x \vec{J}_\mu \cdot \vec{G}_\mu}$$
$$\left(\int D\psi \cdot D\bar{\psi} e^{-\left(\int d^2x (\bar{\psi} \not{D}(G_\mu)\psi + \bar{\eta}\psi + \bar{\psi}\eta)(x)\right)}\right) \quad \text{(D-1)}$$

where $\not{D}(G_\mu) = i\gamma_\mu(\partial_\mu - iG_\mu)$ denotes the self-adjoint Dirac operator in the presence of the external gauge field G_μ and the tensor field strength is given by $F_{\mu\nu} = \partial_\mu G_\nu - \partial\nu G_\mu + i[G_\mu, G_\nu]$.

The hermitean γ-matrices we are using satisfy the (Euclidean) relations.

$$\{\gamma_\mu, \gamma_\nu\} = 2\delta_{\mu\nu}; \quad \gamma_\mu\gamma_5 = i\varepsilon_{\mu\nu}\gamma_\nu; \quad \gamma_5 = i\gamma_0\gamma_1$$

$$\varepsilon_{01} = -\varepsilon_{10} \qquad (\mu, \nu = 0, 1) \quad \text{(D-2)}$$

Tho functional measures in (D-1) are normalized to unity and the fermion measure $D\psi D\bar{\psi}$ is defined in terms of the eigenvalues of the self-adjoInt Dirac operator $\not{D}(G_\mu)$ which insures automatically its gauge invariance.

Our plan to study (D-1) is to implement a convenient change of variables in the fermionic sector of (D-1) in order to get an effective generating functional were the fermion fields are decoupled from the gauge field G_μ ([1], [2]). For this analysis, we are going to use a general decomposition of the gauge field G_μ due to Roskies ([1]) and this will be explained in the following.

Roskies in Ref. [1], has shown that for any gauge field configuration $G_\mu(x)$, there is a unique unitary matrix $\Omega(x)$ taking values in $SU(2)$ and a hermitean matrix $V(x) = e^{-\gamma_5 \vec{\psi}(x)\vec{J}}$ taking value over the axial gauge group $SU(2)$ (whose Lie algebra is generated by the hermitean generators

$\gamma_5 \vec{J}$, with $\vec{J}$ denoting the usual $SU(2)$-generators) such that

$$\begin{aligned} i\gamma\mu \vec{G}_\mu(x)\vec{J} &= \gamma_\mu \partial_\mu (V^{-1}\Omega^{-1})(x)(\Omega V)(x) \\ &= -\gamma_\mu (V^{-1}\Omega^{-1})(x)\partial_\mu (\Omega V)(x). \end{aligned} \tag{D-3}$$

The proof of the validity of the decomposition (D-3) can be accomplished by considering the $j = -i\gamma_5$ complexification of space-time, which is denoted by $\mathbb{C}$ ($\mathbb{C} = \{z = (x_0, x_1) | z = x_0 + jx_1; \bar{z} = x_0 - jx_1$ and $(x_0, x_1) \subset \mathbb{R}^2\}$). In this $\mathbb{C}$ space-time, we can re-write the partial differential equation (D-3) into a single ordinary differential equation ([1])

$$\bar{G} \cdot \vec{\tau} = -2i(\partial_{\bar{z}} W)W^{-1} \tag{D-4}$$

where $\bar{G}$ is the j-complexification of $G_\mu (\bar{G} = G_0 + jG_1)$ and W is an element of $SL(2, \mathbb{C})$ (the associated j-complexification of $SU(2)$).

The equation (D-4) is just the equation for a holomorphic. Principal bundle over $\mathbb{C}$, and, as is well known from differential topology, all such bundles are trivial, which means that a unique global solution W for (D-4) exists.

In order to determine explicitly this solution W, we note that (D-4) can be easily integrated, leading to the result

$$W((x_0, x_1)) = \mathbb{P}\left\{e^{-2i\int_{-\infty}^{z} (\bar{C}\vec{\tau})d\bar{z}}\right\} \tag{D-5}$$

where $d\bar{z} = dx_0 - jdx_1$ and the $SU(2)$ path-ordered integral in (D-5) is taken over the (infinite) straight segment joining the $(-\infty)$ point to the $z = (x_0, x_1)$ point.

By introducing the axial gauge field

$$*\vec{G}_\mu \cdot \vec{\tau} = (\varepsilon_{\mu\alpha}\vec{G}_\alpha) \cdot \vec{\tau} \tag{D-6}$$

we can re-write Eq. (D-5) in the more transparent form

$$W((x_0, x_1)) = \mathbb{P}\left\{e^{\left(-2i\int_{(-\infty)}^{(x_0,x_1)} (\vec{G}_\mu \cdot \vec{\tau} dx_\mu - 2\gamma_5 \int_{(-\infty)}^{(x_0,x_1)} (*G_\mu \cdot \vec{\tau})dx_\mu\right)}\right\} \tag{D-7}$$

where, again, the $SU(2)$ path ordered integral in (D-7) is taken over the straight segment joining the $(-\infty)$ point to the (x_0, x_1) point.

Continuing our study, we can see that the Dirac operator $\not{D}(G_\mu)$ can be re-written in the suitable form ([3], [5])

$$\not{D}(G_\mu) = (\Omega V^{-1})(x)(i\gamma_\mu \partial_\mu)(V^{-1}\Omega^{-1})(x). \tag{D-8}$$

Here, the matrices $\Omega(x)$ and $V(x)$ are respectively the unitary and hermitean factors of the $SL(2, \mathbb{C})$ Wu-Yang factor (D-7).

In order to decouple the fermion fields from the gauge field, we follow Ref. ([1]) by making the variable change.

$$\begin{aligned}\psi(x) &= (\Omega \cdot V)(x)\chi(x) \\ \bar{\psi}(x) &= \bar{\chi}(x)(V \cdot \Omega^{-1})(x)\end{aligned} \tag{D-9}$$

which yields the fermionic generating functional

$$\begin{aligned}\tilde{Z}(\eta, \bar{\eta}] = \int D[x]D[\bar{x}]J[G_\mu]e^{-(\int d^2x(\bar{\chi}(i\gamma_\mu\partial_\mu)\chi} \\ + \bar{\eta}(\Omega \cdot V)\chi + \bar{\chi}V\Omega^{-1}\eta)(x)\end{aligned} \tag{D-10}$$

where the quantum aspect of the variable change (D-9) is taken into account by considering the associated jabobian $J[G_\mu]$ ([1], [2]).

Now, it is important to note that the jacobian $J[G_\mu]$ is given by the ration

$$J[G_\mu] = \frac{\det(\not{D}(G_\mu))}{\det(i\gamma_\mu\partial_\mu)}. \tag{D-11}$$

In order to evaluate the functional fermionic determinant in (D-11), we introduce a family of gauge fields $G_\mu^{(\sigma)}$ $(0 \leq \sigma \leq 1)$ interpolating continuously the zero field configuration $G_\mu^{(\sigma=0)} \equiv 0$ to the considered configuration $G_\mu^{(\sigma=1)} = G_\mu$ in (D-11) and defined by the relation (see Eq. (D-3))

$$i\gamma_\mu G_\mu^{(\sigma)} = -\gamma_\mu(e^{-\sigma\gamma_5\vec{\phi}\vec{\tau}}\Omega^{-1})\partial_\mu(\Omega e^{\sigma\gamma_5\vec{\phi}\vec{\tau}}). \tag{D-12}$$

We note that we are assuming implicitly that we are computing the Jacobian $J[G_\mu]$ in the trivial topological sector of the manifold of the gauge fields configurations, since G_μ is in the same homotopical class of the zero field configuration. As a consequence of this fact, we do not taken into account the zero-modes of the operator $\not{D}(G_\mu)$ in what follows.

Now, it seems important to remark that to evaluate $\det(\not{D}(G_\mu^{(\sigma)}))$ we can consider solely the "reduced" operator

$$\tilde{\not{D}}(G_\mu^{(\sigma)}) = e^{\sigma\gamma_5\vec{\phi}\vec{J}}(i\gamma_\mu\partial_\mu)e^{\sigma\gamma_5\vec{\phi}\vec{J}} \tag{D-13}$$

since $\tilde{\not{D}}(G_\mu^{(\sigma)})$ is related to $\not{D}(G_\mu^{(\sigma)})$ by a similarity transformation defined by the unitary matrix $\Omega(x)$ (see Eq. (D-8)). This result is directly related to the gauge invariance of the Jacobian $J[G_\mu]$, i.e. only the axial $SU(2)$ matrix $V(x)$ contributes to $J[G_\mu]$.

In order to evaluate $\det(\not{D}(G_\mu^{(\sigma)}))$ we proceed as in ([3], [5]). Using the proper-time method to define the functional determinant and making use of the relation

$$\frac{d}{d\sigma}\not{D}(G_\mu^{(\sigma)}) = \gamma^5\vec{\phi}\vec{\tau}\not{D}(G_\mu^{(\sigma)}) + \not{D}(G_\mu^{(\sigma)})\gamma^5\vec{\phi}\vec{\tau} \tag{D-14}$$

we get the following ordinary differential equation

$$\frac{d}{d\sigma}(\log\det(\not{D}(G_\mu^{(\sigma)}))^2) = \lim_{\varepsilon\to 0^+} 4\int d^2x Tr^{(e,D)}$$
$$(\gamma^5(\vec{\phi}\vec{\tau})(x)\langle x|e^{-\varepsilon(\not{D}(G_\mu^{(\sigma)})^2)}|x\rangle \quad \text{(D-15)}$$

where $Tr^{(C,D)}$ denotes the trace over the Dirac and the color indices.

The asymptotic expansion of the operator

$$\langle x|e^{-\varepsilon(\not{D}(G_\mu^{(\sigma)}))^2}|x\rangle = \langle|e^{-\varepsilon\{(-(\partial_\mu - iG_\mu^{(\sigma)})^2 + \frac{i}{2}\varepsilon_{\mu\nu}\gamma_5 F_{\mu\nu}(-iG_\mu^{(\sigma)})\}}|x\rangle$$

is tabulated ([6])

$$\lim_{\varepsilon\to 0^+}\langle x|e^{-\varepsilon(\not{D}(G_\mu^{(\sigma)})^2)}|x\rangle$$
$$= \lim_{\varepsilon\to 0^+}\frac{1}{4\pi\varepsilon}(1-\varepsilon(\frac{iG_{\mu\nu}\gamma_5}{2}F_{\mu\nu}(-iG_\mu^{(\sigma)})))(x). \quad \text{(D-16)}$$

Substituting (D-16) into (D-15), we get the result

$$J[G_\mu] = e^{-\frac{i}{2\pi}\varepsilon_{\mu\nu}\{\int_0^1 d\sigma(\int d^2x Tr^{(c)}(\vec{\phi}\vec{\tau}F_{\mu\nu}(-iG_\mu^{(\sigma)}))(x))\}} \quad \text{(D-17)}$$

We remark that the result (D-17) coincides with the result obtained by Roskies ([1]), and so the σ integration can be done explicitly producing the expression

$$J[G_\mu] = e^{\frac{\varepsilon_{\mu\nu}}{\pi}\{\int d^2x\left(\vec{\phi}\vec{F}_{\mu\nu}(G_\mu)\left(\frac{1}{|\vec{\phi}|\tanh|\vec{\phi}|} - \frac{1}{\sinh^2|\vec{\phi}|}\right)\right)(x)\}} \quad \text{(D-18)}$$

We also note that by considering the vector and axial components of the gauge field $G_\mu^{(\sigma)}$ ($G_\mu^{(\sigma)} = iV_\mu^{(\sigma)} + \varepsilon_{\mu\nu}A_\nu^{(\sigma)}$), we re-obtain the result established in Refs. ([3]) and ([5]).

Finally, the effective generating functional for the model where the fermion fields are decoupled from the gauge fields can be written

$$Z[J_\mu,\beta,\bar{\beta}] = \int D[G_\mu]e^{-\frac{1}{4g^2}\int Tr^{(c)}(F_{\mu\nu}^2)(x)d^2x}\int D^{\text{Haar}}[\Omega]J[G_\mu^\Omega]$$
$$e^{\int(\vec{J}_\mu\vec{G}_\mu^\Omega)(x)d^2x}\left(\int D\chi D\bar{\chi}\, e^{-\{\int d^2x(\bar{\chi}(i\gamma_\mu\partial_\mu)\chi}\right.$$
$$+\bar{\eta}(\Omega V)\chi + \bar{\chi}(V\Omega^{-1})\eta)(x)\} \quad \text{(D-19)}$$

We remark that we have to fix a gauge in (D-19). This gauge is not necessarily the Roskic's gauge ([1], [2]), which choice, will imply to consider $\Omega(x) = \pi$ in (D-19).

From (D-18) we see that the analysis of the fermionic correlation functions are reduced to the computation of the interaction among the $SL(2,\mathbb{C})$

Wu-Yang factors (D-7) with the quantum average defined by the local effective gluonic action (in the Roskies gauge $\Omega \equiv \mathbf{1}$)[2]

$$S^{EFF}[G_\mu] = -\frac{1}{4g^2}\int d^2x Tr^{(c)}(F^2_{\mu\nu})(x) + \ell g J[G_\mu]. \tag{D-20}$$

For instance, the two point fermionic correlation function is given by

$$\langle\psi(x)\bar{\psi}(y)\rangle = \frac{1}{2\pi}\gamma_\mu \frac{x_\mu - y_\mu}{|x-y|^2}\langle W(x)W^{-1}(y)\rangle_{EFF} \tag{D-21}$$

where $\langle\ \ \rangle_{EFF}$ is the quantum average defined by the action (D-20); $W((z)) = W((z_0, z_1))$ is given by Eq. (D-7) and $\frac{1}{2\pi}\gamma_\mu\frac{x_\mu-y_\mu}{|x-y|^2}$ is the free fermion propagator.

2.12 References

[1] R. Roskies, in Festschrift for Feza Gursey's 60th Birthday, (1982) (unpublished).

[2] R.E. Gamboa Saravi, F.A. Schaposnik and J.E. Solomin – Phys. Rev. D – vol. 30, nº 6 (1984).

[3] Luiz C.L. Botelho and M.A. Rego Monteiro, Phys. Rev. D - vol. 30, nº 10 (1984); 2242 Luiz C.L. Botelho, PHD thesis, Centro Brasileiro de Pesquisas Fsicas (1984).

[4] H. Gravert, Math. Annalen 135, 266 (1958).

[5] Orlando Alvares - Nucl. Phys. B238, 61 (1984).

[6] V.N. Romanov and A.S. Schwartz - Teor. Mat. Fiz. 41 (1975), 190.

[7] Robert J. Kares and Myron Bander-Phys. Lett. 96B nº 3,4 (1980), 326.

[2]It appears to be an open problem in the subject the full evaluation of the Non-Abelian Dirac Determinant defined on a Riemman Surface (the Hodge theorem for non abelian connections).

2.13 Appendix E: Path-Integral Bosonization for the Abelian Thirring Model on a Riemann Surface – The $QCD(SU(N))$ String

We study the path integral chiral bosonization of the Abelian-Thirring model defined on a boundaryless Riemann surface as a two-dimensional space-time. This short of 2D fermionic models on random surfaces are expected to be the $QCD(SU(N))$ string – see Chapter 5.

Analysis of quantum field models defined on Riemann surface as two-dimensional space-time is a fundamental issue for strings field theory in Polyakov's approach [1,2].

It is the purpose of the letter to solve exactly the Abelian-Thirring model defined on a Riemann surface in the Framework of chiral path integrals [3].

We start our analysis by considering the Abelian-Thirring model associated to a complex spin field associated to a complex structure (θ^i, ϕ^i) of a genus g Riemann surface $D^{(g)}$

$$\mathcal{L}(\psi, \bar{\psi}))(\theta^i, \phi^i) = \bar{\psi} i\gamma^\mu D_\mu \psi + \frac{c^2}{2}(\bar{\psi}\gamma^\mu\psi)^2. \tag{E-1}$$

Here thd Dirac operator is given by

$$i\gamma^\mu D_\mu = i\gamma \hat{e}^\mu_a \partial_\mu \tag{E-2}$$

where $\hat{e}^\mu_a$ are fixed background two-being satisfying the topological genus constraint

$$\int_{D^{(g)}} \sqrt{\hat{g}}\, R(\hat{g}) = 2\pi(2-2g). \tag{E-3}$$

$R(\hat{g})$ is the scalar of curvature associated to $\hat{g}_{\mu\nu}$ and $\omega_{\mu ab}(\hat{g})$ is the spin connection defined by the relation $\nabla_\mu \partial_\nu = 0$.

The $\gamma^\mu = \hat{e}^\mu_a \gamma_a$ Eucidean (curved) Dirac matrices are defined by the relationship below ($\xi \in D^{(g)}$)

$$\begin{cases} \{\gamma_\mu, \gamma_\nu) + (\xi) = 2\hat{g}_{\mu\nu}(\xi), \\ \gamma^\mu(\xi)\gamma_5 = i\left(\frac{\varepsilon^{\mu\nu}\gamma_\nu}{\sqrt{\hat{g}}}\right)(\xi), \end{cases} \tag{E-4}$$

where γ_a are the usual flat-space Dirac matrices.

In the framework of path integrals, the generating functional of the Green's function of the (mathematical) quantum field theory associated with the Lagrangian eq. (E-1) is defined by the following covariant functional integration [2]

$$Z[\rho, \bar{\rho}] = \int d^c[\psi] d^c[\bar{\psi}] \times \exp\left[-\int_{D^{(g)}} d^2\xi (\sqrt{\hat{g}}\, \mathcal{L}(\psi, \bar{\psi}))(\xi)\right]$$
$$\times \exp\left[-\int_{D^{(g)}} d^2\xi (\sqrt{\hat{g}}\, (\bar{\rho}\psi + \bar{\psi}\rho))(\xi)\right]. \tag{E-5}$$

It is worth pointing out that the classical action in eq. (E-5) s invariant under the local diffeomorphism group and the global Abelian and Abelian-chiral groups acting on the spin field restrict to any local region R of $D^{(g)}$. These symmetries have the associated Noether conserved currents

$$\nabla_\mu(\bar{\psi}\gamma^5\gamma^\mu\psi) = 0; \quad \nabla_\mu(\bar{\psi}\gamma^\mu\psi) = 0. \tag{E-6}$$

In order to implement the path-integral gauge and local diffeomorphism invariant bosonization, we rewrite the fermion interaction term in the Hubbard-Stratonovitch form by using an ausiliary vector field $A_\mu(\xi)$

$$\begin{aligned} Z[\rho,\bar{\rho}] = \int & d^c[\psi]d^c[\bar{\psi}]d^c[A_\mu] \\ & \times \exp\left[-\int_{D^{(g)}} d^2\xi(\sqrt{\hat{g}}\left[\bar{\psi}i\gamma^\mu(d_\mu + cA_\mu)\psi + \frac{1}{2}A_\mu A^\mu\right](\xi)\right]\Big] \\ & \times \exp\left[-\int_{D^{(g)}} d^2\xi\sqrt{\hat{g}}\,(\bar{\rho}\psi + \bar{\psi}\rho)(\xi)\right]. \end{aligned} \tag{E-7}$$

Let us now proceed as in [4–6] by making the local field change in eq. (E-7)

$$A_\mu(\xi) = -\left(\frac{\varepsilon^{\mu\nu}}{\sqrt{\hat{g}}}\partial_\nu\eta\right)(\xi) + A_\mu^H(\xi), \tag{E-8}$$

$$\psi(\xi) = (\exp[i\gamma_5\eta(\xi)])\cdot\chi(\xi), \tag{E-9}$$

$$\bar{\psi}(\xi) = \bar{\chi}(\xi)\cdot\exp[i\gamma_5\eta(\xi)], \tag{E-10}$$

where $\nabla^\mu(A_\mu - A_\mu^H) \equiv 0$ and $A_\mu^H(\xi)$ is the Hodge topological vector field which is explicitly given in terms of canonical Abelian differentials ω_i nd their complex conjugates $\bar{\omega}_i$ [7]

$$A_\mu^H(\xi) = 2\pi\sum_{l=1}^{g}(p_i\cdot\alpha_\mu^i(\xi) + r_i\beta_\mu^i(\xi)), \tag{E-11}$$

$$\alpha_\mu^i(\xi) = -\bar{\Omega}_{ik}(\Omega-\bar{\Omega})_{kj}^{-1}(\xi) + c.c., \tag{E-12}$$

$$\beta_\mu^i(\xi) = (\Omega - \bar{\Omega}_{ij}^{-1}\omega_\mu^i(\xi) + c.c. \tag{E-13}$$

The period matrix Ω is defined by

$$\int a^i = \delta_{ij}; \int_{b^i} a^j = \Omega_{ij}, \tag{E-14}$$

where a^i and b^i are (canonical) homology cycles on $D^{(g)}$.

As it has been shown by Fujikawa [5], the transformation of eqs. (E-9)–(E-10) are not free of cost, since the functional measures $d^c[\psi]d^c[\bar{\psi}]$ are defined in terms of the normalized eigenvectors of the covariant and $U(1)$ gauge invariant Dirac operator eq. (E-2) in the presence of the auxiliary vector field A_μ.

The associated Jacobian of eqs. (E-9), (E-10) is given by [6]

$$d^c[\psi]d^c[\psi] = d^c[\chi]d^c[\bar{\chi}] \times \frac{\det[i\gamma^\mu(D_\mu + cA_\mu)]}{\det[i\gamma^\mu(D_\mu + cA_\mu^H)]}. \tag{E-15}$$

At this point we note that after the chiral change takes place the new quantum fermionic vacuum is defined by the fermionic field $\chi(\xi)$ (with the same spin structure of $\psi(\xi)$) in the presence solely of the Hodge topological field A_μ^H (eq. (E-1)).

The Jacobian associated to eq. (E-8) is [7]

$$d^c[A_\mu] = d^c[\eta]\left((2\pi)^{2g}\prod_{l=1}^{g} dp_i dr_i\right) \times \det{}^{1/2}\begin{pmatrix} \langle\alpha_\mu^i, \alpha_\mu^i\rangle & \langle\alpha_\mu^i, \beta_\mu^i\rangle \\ \langle\beta_\mu^i, \alpha_\mu^i\rangle & \langle\beta_\mu^i, \beta_\mu^i\rangle \end{pmatrix}, \tag{E-16}$$

where the covariant scalar product in the space of vector fields in $D^{(g)}$ is defined by

$$\langle\Sigma_\mu, \theta_\mu\rangle = \int_{D^{(g)}} d^2\xi(\sqrt{\hat{g}}\,\hat{g}^{\alpha\beta}\Sigma_\alpha\theta_\beta)(\xi). \tag{E-17}$$

Let us remark that with this definition we have

$$\langle\omega_\mu^i, \omega_\mu^j\rangle = 2\mathrm{Im}\Omega_{ij}. \tag{E-18}$$

So, we face the problem of the evaluation of the ration of two Dirac determinants related themselves by a chiral rotation

$$J[A_\mu] = \frac{\det[\exp[ic\gamma_5\eta]i\gamma^\mu(D_\mu + cA_\mu)\exp[ic\gamma_5\eta]]}{\det[i\gamma^\mu(D_\mu + cA_\mu^H)]}. \tag{E-19}$$

Byfollowing the procedure of ref.[6] we, at first, introduce a One-parameter family of Dirac operators interpolating the Dirac operator $i\gamma^\mu(D_\mu + cA_\mu^H) = \mathbb{D}(A_\mu^H)$ and the chirally rotated $\exp[ic\gamma_5\eta)\mathbb{D}(A_\mu^H)\cdot\exp[ic\gamma_5\eta]$

$$\mathbb{D}^{(\zeta)}(A_\mu) = \exp[ic\gamma_5\zeta\eta]\mathbb{D}(A_\mu^H)\cdot\exp[ic\gamma_5\zeta\eta], \quad (0 \le \zeta \le 1). \tag{E-20}$$

By using a proper-time prescription to define the functional determinant of $\mathbb{D}^{(\zeta)}$ (after making the analytic extension $c = -i\bar{c}$), we have the following differential equation for $\log\det\mathbb{D}^{(\zeta)}$

$$\begin{aligned}\frac{d}{d\zeta}\log\det\mathbb{D}^{(\zeta)} &= -2\lim_{\varepsilon\to0^+} Tr[\bar{c}\gamma_5\eta\exp[-\sigma\mathbb{D}^{(\zeta)^2}]\times(1-\mathbb{P}^{(\zeta)})]_{\varepsilon}^{1/\varepsilon}\\ &+\lim_{\varepsilon\to0^+}\int_{\varepsilon}^{1/\varepsilon}\frac{d\sigma}{\sigma}Tr\left(\exp[-\sigma\mathbb{D}^{(\zeta)^2}]\frac{d}{d\zeta}\mathbb{P}^{(\zeta)}\right)\\ &= I_{(1)}^{(\zeta)}[A_\mu]+I_{(0)}^{(\zeta)}[A_\mu],\end{aligned} \tag{E-21}$$

where $\mathbb{P}^{(\zeta)} = \sum_n\langle\,,\phi_n^{(0),(\zeta)}\rangle\phi_n^{(0),(\zeta)}$ denotes the projection over the zero modes $\phi_n^{(0),(\zeta)}$ of the Dirac interpolating operator $\mathbb{D}^{(\zeta)}$. These zero modes are related by an analytically continued chiral rotation to those of $\mathbb{D}(A_\mu^H)$

$$\phi_n^{(0),(\zeta)} = \exp[-\bar{c}\gamma_5\zeta\eta]\cdot\bar{\phi}_n^{(0)} \tag{E-22}$$

and

$$\mathbb{D}(A_\mu^H)\cdot\bar{\phi}_n^{(0)} = 0. \tag{E-23}$$

Since $\mathbb{D}^{(\sigma)^2}(A_\mu)$ is a self-adjoint invertible operator in the manifold orthogonal to the subspace generated by the zero modes, we can use the Seeley-De Witt technique to evaluate the first term in eq. (E-21) which yields

$$\begin{aligned}I_{(1)}^{(\zeta)}[A_\mu] &= \lim_{\varepsilon\to0^+} Tr[\bar{c}\gamma_5\eta\exp[-\sigma\mathbb{D}^{(\zeta)^2}](1-\mathbb{P}^{(\zeta)})]|_{\varepsilon}^{1/\varepsilon}\\ &= -\frac{2}{\pi}\zeta Tr\left[-i\bar{c}\left(\eta\frac{1}{\sqrt{\hat{g}}}\partial_\alpha(\hat{g}^{\alpha\beta}\partial_\beta)\eta\right)+\frac{\varepsilon_{\mu\nu}}{2}F^{\mu\nu}(A^H)\right].\end{aligned} \tag{E-24}$$

The second term on the left side of eq. (E-21) is easily evaluated giving the result

$$\begin{aligned}I_{(1)}^{(\zeta)}[A_\mu] &= \lim_{\varepsilon\to0^+}\int_{\varepsilon}^{1/\varepsilon}\frac{d\sigma}{\sigma}Tr\left(\exp[-\sigma\mathbb{D}^{(\zeta)^2}]\frac{d}{d\zeta}\mathbb{P}^{(\zeta)}\right)\\ &= \lim_{\varepsilon\to0^+}(4\log\varepsilon\cdot\bar{c})\sum_n\int_{D(g)}d^2\xi\left(\sqrt{\hat{g}}\,\bar{\phi}_n^{(0)}\cdot\eta\bar{\bar{\phi}}_n^{(0)}\right)(\xi).\end{aligned} \tag{E-25}$$

The final result for the functional determinants ratio eq. (E-19) is thus given by

$$\begin{aligned}J[A_\mu] &= \frac{(C^{(R)})^2}{\pi}\int_{D(g)}d^2\xi\frac{1}{2}\left(\sqrt{\hat{g}}\,\partial_\alpha\eta\hat{g}^{\alpha\beta}\partial_\beta\eta\right)(\xi)\\ &+\frac{C^{(R)}}{\pi}\int_{D(g)}d^2\xi\left[(\varepsilon_{\mu\nu}F^{\mu\nu}(A_\mu^H)\cdot\eta)\sqrt{\hat{g}}\right](\xi)\\ &- ic^R\sum_n\int_{D(g)}d^2\xi\left(\sqrt{\hat{g}}\,\bar{\phi}_n^{(0)}\cdot\eta\bar{\bar{\phi}}_n^{(0)}\right)(\xi),\end{aligned} \tag{E-26}$$

here $c^{(R)}$ is the usual multiplicative infrared coupling constant renormalization due to zero-mode terms.

This result generalizes and produces a proof for the conjectured functional determinant of ref. [7] which was deduced in the particular case of nonexistence of zero modes.

The generating functional thus takes the more invariant form

$$Z[\rho,\bar{\rho}] = \int dm(p_i,r_i) Z^{(0)}[\rho,\bar{\rho},(p_i,r_i)], \tag{E-27}$$

where the measure over the (p_i, r_i) parameters is given by [7]

$$\begin{aligned} dm(p_i,r_i) &= (2\pi)^{2g} \prod_{l=1}^{g} dp_l \cdot dr_l \times \det \begin{pmatrix} \langle \alpha^i_\mu, \alpha^i_\mu \rangle & \langle \alpha^i_\mu, \beta^j_\mu \\ \langle \beta^i_\mu, \alpha^j_\mu \rangle & \langle \beta^i_\mu, \beta^i_\mu \rangle \end{pmatrix} \\ &\quad \times \exp\Big[- 2\pi^2 \int_{D(g)} d^2\xi \Big\{ \sqrt{\hat{g}}\,[(p_k \bar{\Omega}_{ki} - r_i) \\ &\quad \times (\Omega)^{-1}_{ij} (\Omega_{jl} p_l - r_i)] \Big\}(\xi) \Big]. \end{aligned} \tag{E-28}$$

The (bosonized) generating functional is explicitly given by

$$\begin{aligned} Z^{())}[\rho,\bar{\rho}] &= \int d^c[\eta] \exp\big[iW[\hat{\phi}^{(0)}_n, \bar{\hat{\phi}}^{(0)}_n, A^H_\mu]\big] \\ &\quad \times \int d^c[\chi] d^c[\bar{\chi}] \exp\Big[-\frac{1}{2}\Big(1 - \frac{c^{(R)^2}}{\pi}\Big) \int_{D(g)} d^2\xi \sqrt{\hat{g}} [\hat{g}^{\alpha\beta} \partial_\alpha \eta \partial_\beta \eta)(\xi) \\ &\quad + (\bar{\chi} i \gamma^\mu (D_\mu + c^{(R)} A^H_\mu) \chi)(\xi) \\ &\quad + (\bar{\chi} \exp[ic^{(R)}\gamma_5 \eta] \rho + \bar{\rho} \exp[ic^{(R)} \gamma_5 \eta] \chi)(\xi) \Big], \end{aligned} \tag{E-29}$$

where the functional $W[\hat{\phi}^{(0)}_n, \bar{\hat{\phi}}^{(0)}_n, A^H_\mu]$ is defined by the interaction with the (external) zero-mode fermion fields $\hat{\phi}^{(0)}_n, \bar{\hat{\phi}}^{(0)}_n$

$$\begin{aligned} W[\hat{\phi}^{(0)}_n, \bar{\hat{\phi}}^{(0)}_n, A^H_\mu] &= \int_{D(g)} d^2\xi \sqrt{\hat{g}} \Big[\Big(- i \frac{c^{(R)}}{\pi} \varepsilon_{\mu\nu} F^{\mu\nu}(A^H) \eta \Big) \\ &\quad + (-c^{(R)} \hat{\phi}^{(0)}_n \eta \bar{\hat{\phi}}^{(0)}_n) \Big](\xi). \end{aligned} \tag{E-30}$$

We remark that the fermions $\chi(\xi)$ still interact with the Hodge topological field A^H_μ by the minimal gauge invariant interaction $\mathbb{D}(A^H_\mu)$ and with the $\eta(\xi)$ field by the coupling with the source term.

Let us exemplify our main result, eq. (E-29), by displaying the general structure of the two-point fermion correlation function

$$\langle \psi_\alpha(\xi_1)\bar{\psi}_\beta(\xi_2)\rangle = \frac{1}{Z[\rho,\tilde{\rho}]}\int_{-\infty}^{+\infty} dm(p_i,r_i)\det[i\gamma^\mu(D_\mu + c^{(R)}A_\mu^H)] \times \exp\left[-\frac{1}{2}\frac{c^{(R)^2}}{1-(c^{(R)^2}/\pi)}\Delta^{-1}(\xi_1,\xi_2)\right]\mathbb{D}^{-1}(A_\mu^H), \tag{E-31}$$

where $\Delta^{-1}(\xi_1,\xi_2)$ is the Green's function of the Laplace operator on the Riemann surface $D^{(g)}$ and $\mathbb{D}^{-1}(A_\mu^H) = (i\gamma^\mu(D_\mu + c^{(R)}A_\mu^H))^{-1}(\xi_1,\xi_2)$ is the Gree's function of the Dirac operator with spin structure (θ_i,ϕ_i) in the presence of the topological Hodge vector field $A_\mu^H(\xi)$[1].

The determinant in eq. (E-31) was exactly evaluated ref.[1] and expressed in terms of ϑ-functions

$$\det i\gamma^\mu(D_\mu + c^{(R)}A_\mu^H) = |l(\Omega)|^2\cdot\left|\vartheta\begin{bmatrix}\frac{1}{2}+\theta^i\\ \frac{1}{2}-\phi^i\end{bmatrix}(0|\Omega)\right|. \tag{E-32}$$

The Green's function of the Laplace operator may be expressed in terms of the theta-functions

$$\Delta^{-1}(\xi_1,\xi_2) = -\frac{1}{4\pi}\log|\vartheta[(\xi_1|\Omega| - \vartheta[(\xi_2|\Omega)]| + \frac{\Im(\xi_1-\xi_2)^2}{\det(\Im\Omega)}. \tag{E-33}$$

Finally a formal expression for the Green's function of the Dirac operator is given by [3]

$$\exp\left[-i\frac{c^{(R)}}{2}\int_{C_{\xi_1,\xi_2}}(A_\mu^H + \gamma_5\varepsilon_{\mu\nu}A^{\nu,H})d\xi^\mu\right]\times(i\gamma^\mu D_\mu(A^H))^{-1}_{(\phi^i,\theta^i)}(\xi_1,\xi_2) \times\exp\left[-i\frac{c^{(R)}}{2}\int_{C_{\xi_1,\xi_2}}(A_\mu^H + \gamma_5\varepsilon_{\mu\nu}A^{\nu,H})d\xi^\mu\right] \tag{E-34}$$

where C_{ξ_1,ξ_2} is an arbitrary contour on the Riemann surface $D^{(g)}$ which has a nonempty intersection with each canonical homology cycles on $D^{(g)}$ and connecting the points ξ_1 and ξ_2.

As we have shown, chiral changes in path integrals even for fermion model on a Riemann surface provid a quick, mathematically and conceptually simple way to analyse these models.

2.14 References

[1] Alvares-Guam L., Moore G. and Vafa C., *Commun. Math. Phys.*, 106 (1980) 1.
[2] Belavin A.A. and Knizhnic V.G., *Sov. Phys. JETP*, 64 (1986) 214.
[3] Botelho Luiz C.L., *Phys. Rev.* D, 34 (1986) 3250.
[4] Botelho Luiz C.L., *Phys. Rev.* D, 35 (1987) 1515.
[5] Fujikawa K., *Phys. Rev.*, D, 21 (1980) 2848.
[6] Botelho Luiz C.L., *Phys. Rev.* D, 39 (1989) 3051.
[7] Freedman D.Z. and Pilch K., *Phys. Lett.* B, 213 (1988) 331.

Remark: For a random surfaqce $\{X^\mu(\xi)\}$ background on the proper-time (light-cone) gauge

$$(\hat{e}^a_\mu \hat{e}^b_\nu)(\xi) = \rho(\xi)\delta_{\mu\nu}$$

with

$$\rho(\xi) = (\partial_+ X^\mu)(\partial_- X_\mu)(\xi) \neq 0$$

and

$$(\partial_+ X^\mu)^2 = (\partial_- X^\mu)^2 = 0$$

as in Chapter 2, §2.4 and perhaps added with a Gross-Neven "mass" term $(\bar{\psi}, \psi)^2$, one expects tht the resulting generalized "Elfin" string path integral is the correct QFT definition for QCD (see also ref. [19] on contents page xi).

Chapter 3

Critical String Wave Equations and the $QCD(U(N_c))$ String

We present a simple non mathematical rigorous proof that self-avoiding fermionic strings solutions solve formally (in a Quantum Mechanical Framework) the $QCD(U(U_c))$ Loop Wave Equations written in terms of random loops for a certain class of self-avoiding surfaces.

3.1 Introduction

We aim, in this chapter to present a formal interacting string solution for the Migdal-Makeenko Loop Wave Equation for the colour group $(U(N_c))$ (Ref. [1] and references therein) for a special class of self-avoiding surfaces.

Our main tool to solve the Migdal-Makeenko Loop Wave Equation is based on the remark made in the Sect. 3.2 of this chapter, where we address tshe problem of solving critical string wave equations by string functional integral by applying simple rules of the operatorial calculus of Quantum Mechanics. We thus apply the results of Sect. 3.2 to present a string functional integral solution for the Migdal-Makeenko Loop Wave Equation for the colour group $U(N_c)$ at this class of special self-intersecting surfaces.

3.2 The Critical Area-Diffusion String Wave Equations

Let us start this section by briefly reviewing our general procedure to write diffusion string wave equations for Bosonic non-critical string [2]. The first step is by considering the following fixed area string propagator in 2D induced quantum gravity string quantization framework.

$$\mathcal{G}[C^{\rm out}, C^{\rm in}, A] = \int D^c[g_{ab}]D^c[X_\mu] \times \delta\left(\int_D d\sigma d\tau\sqrt{g(\sigma,\tau)} - A\right) \times \exp(-I_0(g_{ab}, X_\mu, \mu^2 = 0)). \tag{3.1}$$

Here the string surface parameter domain is taken to be the rectangle $D = \{(\sigma,\tau), -\pi \le a \le \pi, 0 \le \tau \le T\}$. The action $I_0(g_{ab}, X_\mu, .\mu^2 = 0)$ is the Brink-Di Vecchia-Howe covariant action with a zero cosmological term and the covariant functional measures $D^c[g_{ab}]D^c[X_\mu]$ are defined over all cylindrical string world sheets without holes and handles with the initial and finaql string configurations as unique non-trivial boundaries: i.e. $X_\mu(c,0) = C^{\text{in}}$, $X_\mu(\sigma,T) = C^{\text{out}}$.

In order to write an area diffusion wave equation for (3.1), we exploit an identity which relates its area variation (the Mandelstam area derivative for strings) to functional variations on the conformal factor measure when one fixes the string diffeomorphism group in (3.1) by imposing the conformal gauge $g_{ab}(sigma,\tau) = \rho(\sigma,\tau)\delta_{ab}$ (see Refs. [1-3]). This procedure yields, thusm the following area diffusion string Euclidean wave equation

$$\frac{\partial}{\partial A}\mathcal{G}[C^{\text{out}}, C^{\text{in}}, A = \int_{-\pi}^{\pi} d\sigma \left(-\frac{\delta^2}{2e^2_{\text{in}}(\sigma)\delta C^{\text{in}}_\mu(\sigma)\delta C^{\text{in}}_\mu(\sigma)} + \frac{1}{2}C^{\text{in}}_\mu(\sigma)^2 \right.$$
$$\left. + \frac{26-D}{24\pi} \lim_{r\to 0^+} [R(\rho(\sigma,\tau)) + C_\infty] \right) \times \mathcal{G}[C^{\text{out}}, C^{\text{in}}, A]. \tag{3.2}$$

At this point a subtle difficulty appears when the theory describged by (3.1) is at its critical dimension $D = 26$ since the conformal field $\rho(\sigma,\tau)$ decouples from the theory, making it subtle to implement the fixed area constrint in (3.1). It is instructive to point out that for a cylinder surface without holes and handles with non trivial boundaries, the argument that the fixed area constrint is simply fixing the modulus λ of the (torus) conformal gauge $g_{ab}(\sigma,\tau) = \rho(\sigma,\tau)((d\sigma)^2 + \lambda^2(d\tau)^2)$ is insufficient to cover the case of "string creation" from the vacuum as we will need in Sect. 3. This is because in this case $\lambda = 0$ and the string world sheet still has a non-zero area. Note that the topology of this string world sheet creation process is now a hemisphere which again makes impossible the use of the modulus λ as an area parameter.

However, it makes sense to consider the limit of the parameter $D = 26$ directly in our string diffusion (3.2) which reproduces the usual critical string wave equations ((3.2) with $D = 26$ and $\rho(\sigma,\tau) = 1$).

In this short section we intend to show that the following critical string propagator

$$\mathcal{G}[C^{\text{out}}, C^{\text{in}}, A] = \int D^F[X^\mu(\sigma,\tau)]$$
$$X^\mu(\sigma,0) = C_\mu^{\text{in}}(\sigma), \quad X^\mu(\sigma,A) = C_\mu^{\text{out}}(\sigma)$$
$$\exp\left\{-\frac{1}{2}\int_0^A d\tau \int_{-\pi}^{\pi} d\sigma \left[\left(\frac{\partial X^\mu}{\partial \sigma}\right)^2 + \left(\frac{\partial X^\mu}{\partial \tau}\right)^2\right](\sigma,\tau)\right\} \tag{3.3}$$

where the intrinsic string time parameter A is identified with the area diffusion variable, satisfies the string critical diffusion wave equation.

To show this simple result we evaluate the A-derivative of (3.3) by means of Leibnitz's rule

$$\frac{\partial}{\partial A}\mathcal{G}[C^{\text{out}}, C^{\text{in}}, A]$$
$$= \frac{-1}{2}\left\{\lim_{\tau\to A^-}\left\langle \int_{-\pi}^{\pi} d\sigma \left[\left(\frac{\partial X^\mu}{\partial \sigma}\right)^2 + \left(\frac{\partial X^\mu}{\partial \tau}\right)^2\right](\sigma,\tau)\right\rangle\right\} \tag{3.4}$$

where the surface average $\langle\ \rangle_s$ is defined by Bosonic path-integral in (3.3).

In order to translate the path integral relation (3.4) into a operator statement, we use the usual Heisenberg Commutation Relations for two-dimensional (2D) free fields on D (with the Bidimensional Planck constant = Regge slope parameter set to the value one)

$$[\Pi_\mu(\sigma,\tau), X_\nu(\sigma',\tau)] = i\delta(\sigma-\sigma')\delta_{\mu\nu} \tag{3.5}$$

and its associated Schrdinger representation for $\tau = A$ (that are the quantum mechanical definition of the lopp derivatives operators [1]).

$$\Pi_\mu(\sigma,A) = \lim_{\tau\to A}\left\langle \frac{\partial}{\partial\tau}X^\mu(\sigma,\tau)\right\rangle_s = +i\frac{\delta}{C_\mu^{\text{out}}(\sigma)} \tag{3.6}$$

$$\left|\frac{dC_\mu^{\text{out}}(\sigma)}{d\sigma}\right|^2 = \lim_{\tau\to A}\left(\frac{\partial X^\mu(\sigma,\tau)}{\partial\tau}\right)^2. \tag{3.7}$$

After substituting (3.6)–(3.7) into (3.4) we obtain the desired result

$$\frac{\partial}{\partial A}\mathcal{G}[C^{\text{out}}, C^{\text{in}}, A] = -\left(\int_{-\pi}^{\pi} d\sigma \left(-\frac{\delta^2}{2\delta C_\mu^{\text{out}}(\sigma)\delta C_\mu^{\text{out}}(\sigma)} + \frac{1}{2}|C_\mu^{\text{out}'}(\sigma)|^2\right]\right). \tag{3.8}$$

Let us point out that general string wave functionals (the Schrdinger representation for the theory's quantum states) may be formally expanded in terms of the eigenvunctions of the quantum string Hamiltonian (the string wave operator in (3.8))

$$-\Delta_c \psi_E[c] = -\left\{ \int_{-\pi}^{\pi} d\sigma \left(-\frac{\delta^2}{2\delta C_\mu(\sigma)\delta C_\mu(\sigma)} \frac{1}{2} |C'_\mu(\sigma)|^2 \right) \right\} \psi_E(c) = E\psi_E[c] \tag{3.9}$$

$$\psi[c] = \sum_{\{E\}} \rho(E)\psi_e[c]. \tag{3.10}$$

The functionals endowed with the (formal) inner product given by

$$\langle \psi[c] | \Omega[c] \rangle = \int D^F[c] \cdot \psi^*[c] \cdot \Omega[c] \tag{3.11}$$

constitute a Hilbert space where the string Laplacian $-\Delta_c$ is formally a Hermitian operator.

It is worth remarking that an explicit expression for the Green's Function

$$(-\Delta_c)^{-1}(C^{\text{out}}, C^{\text{in}}) = \sum_{\{E\}} \psi_E[C^{\text{out}}]\psi_E^*[C^{\text{in}}]/E$$

of the string Laplacian in terms of the cylindrical string propagator (3) may be easily obtained.

In order to deduce this expression we integrate both side4 of (3.8) with respect to the A-variable. Considering now the Asymptotic Behaviors.

$$\lim_{A \to \infty} \mathcal{G}[C^{\text{out}}, C^{\text{in}}, A] = 0 \tag{3.12}$$

$$\lim_{A \to 0} \mathcal{G}[C^{\text{out}}, C^{\text{in}}, A] = \delta^F(C^{\text{out}} - C^{\text{in}}) \tag{3.13}$$

we obtain the relationship

$$\Delta^F(C^{\text{out}} - C^{\text{in}}) = -\Delta_c \left(\int_0^\infty dA \mathcal{G}[C^{\text{out}}, C^{\text{in}}, A] \right) \tag{3.14}$$

leading thus to the following identify

$$(-\Delta_c)^{-1}[C^{\text{out}}, C^{\text{in}}] = \left(\int_0^\infty dA \mathcal{G}[C^{\text{out}}, C^{\text{in}}, A] \right) \tag{3.15}$$

3.3 A Bilinear Fermion Coupling on a Self-Interacting Bosonic Random Surface as Solution of $QCD(U(N_c))$ Migdal-Makeenko Loop Equantion

Let us start this section by considering the (non-renormalized) Migdal-makeenko Loop Equation satisfied by the Quantum Wilson Loop in the form of Ref. [4] for the colour group $U(N_c)$.

$$-\Delta_c\langle W_{\ell\ell}[C_{X(-\pi)X(\pi)}\rangle = (g^2 N_c)\int_{-\pi}^{\pi} d\sigma' \frac{dX^\mu(\sigma)}{d\sigma}\cdot\frac{dX^\mu(\sigma')}{d\sigma'}$$
$$\times\,\delta^{(D)}(X_\mu(\sigma)-X_\mu(\sigma')\langle W_{kp}[C_{X(-\pi)X(\sigma)}]W_{p\ell}[C_{X(\sigma)X(\pi)}]\rangle. \quad (3.16)$$

The Quantum Wilson Loop is given by

$$\langle W_{k\ell}[C_{X(-\pi)X(\pi)},A_\mu(x)]\rangle$$
$$=\frac{1}{N_c}\left\langle T_R^{(c)}\left(\exp-\int_{-\pi}^{\pi} d\sigma(A_\mu(X_\mu(\sigma))\cdot X'^\mu(\sigma)\right)\right\rangle_{k\ell}. \quad (3.17)$$

As usual, $A_\mu(x)$ denotes the usual $U(N)$ colour Yang-mills field which possesses an additional, not yet specified intrinsic global "Flavor" group $O(M)$ represented by matrix indices (k,ℓ). The average $\langle\ \rangle$ is given by the $U(N)$-colour Yang-Mills field theory.

Let us consider the following critical non-linear interacting Fermionic String theory first considered in Ref. [5]

$$S[X_\mu(\sigma,\tau),\psi_{(k)}(\sigma,\tau)] = \frac{1}{2}\int_0^A d\tau\int_{-\pi}^{\pi} d\sigma\left[\left(\frac{\partial X^\mu}{\partial\tau}\right)^2+\left(\frac{\partial X^\mu}{\partial\sigma}\right)^2\right](\sigma,\tau)$$
$$+\int_0^A d\tau\int_{-\pi}^{\pi} d\sigma[\bar\psi_{(k)}(\gamma^a\partial_a)\psi_{(k)}](\sigma,\tau)$$
$$+\frac{\beta}{2}\int_0^A d\tau\int_0^A d\tau'\int_{-\pi}^{\pi} d\tau'\int_{-\pi}^{\pi} d\sigma'$$
$$\times\,(\psi_{(k)}\bar\psi_{(k)})(\sigma,\tau)\times T^{\mu\nu}(X_\alpha(\sigma,\tau))\delta^{(D)}$$
$$\times\,(X_\alpha(\sigma,\tau)-X_\alpha(\sigma',\tau'))T_{\mu\nu}(X_\alpha(\sigma',\tau')). \quad (3.18a)$$

The notation is as follows: the string vector position is described by the $2D$-fields $X_\mu(\sigma,\zeta)$ with the Drichlet boundary condition $X_\mu(\sigma,A) = C_{X(-\pi),X(\pi)}$; i.e., our special class of self-intersecting surfaces $S = \{X_\mu(\sigma,\zeta), -\pi\le\sigma\le\pi, 0\le\zeta\le A)$ has a unique boundary the fixed Loop $C_{X(-\pi),X(\pi)}$ of (3.16). The surface orientation tensor where it is defined is given by

$$T_{\mu\nu}(X_\mu(\sigma,\tau)) = \frac{\varepsilon^{ab}}{\sqrt{2}}\frac{\partial_a X^\mu\partial_b X^\nu}{\sqrt{h}} \quad (3.18b)$$

with $h = \det\{h_{ab}\}$ and $h_{ab}(\sigma,\tau) = \partial_a X^\mu \partial_b X^\mu$. Note that S possesses self-intersecting lines such that $X_\mu(\sigma,\tau) = X_\mu(\sigma',\tau')$ with $0 < \{\tau,\tau'\} < A$ has non-trivial self-intersecting lines solutions. For $\tau = A = \tau'$, $X_\mu(\sigma, A) = C_\mu^{\text{out}}(\sigma)$ posseses solely simple isolated self-intersections points (eights loops), however with $T^{\mu\nu}(X_\alpha(\sigma,A))T_{\mu\nu}(X_\beta(\sigma',A)) = 0$ for $\sigma \neq \sigma'$ (the Fermion Exclusion Pauli Principle). Additionally we have introduced a set of single-valued intrinsic Majorana $2D$-spinors on the surface domain parameter $D = \{(\sigma,\tau),\ 0 \leq \tau \leq A; -\pi \leq 0 \leq \tau\}$. They are chosen to belong to a real representation of the flavor group $O(22)$ since for this group we have cancelled exactly the theory's conformal anomaly $(26 = 4 + 22)$, which in turn leads to the vanishing of the kinetic term associated to the conformal factor $\rho(\sigma,\zeta)$ (see Ref. [1]). We further impose as a boundary condition on these Fermions the vanishing of the Fermion energy-tensor projected on the Loop $C_{X(-\pi),X(\pi)}$. Let us point out that the Weill symmetry makes sense to speak in conformal anomaly in our theory (3.16) which preservation at quantum level by its turn will determine the string flavor group to be the "String" Weinberg-Salam group $O(22)$ (see Ref. [1]).

Associated to the non-linear strings theory (3.18) we consider the following Fermionic propagator for a fixed string world sheet $\{X_\mu(\sigma,\tau)\}$

$$\begin{aligned}
&\bar{Z}_{k\ell}[C_{X(-\pi),X(\pi)}; X_\mu(\sigma,\zeta), A] \\
&\quad = \int D^F[\psi_k(\sigma,\tau)](\psi_{(k)}(-\pi,A)\bar{\psi}_{(\ell)}(\pi,A) \\
&\qquad \times \exp\left\{-\int_0^A d\tau \int_{-\pi}^{\pi} d\sigma(\bar{\psi}_{(k)}(\gamma^a\partial_a)\psi_{(k)})(\sigma,\tau)\right\} \\
&\qquad \times \exp\bigg\{-\frac{\beta}{2}\int_0^A d\tau\int_{-\pi}^{\pi} d\sigma \int_0^A d\tau' \int_{-\pi}^{\pi} d\sigma'(\bar{\psi}_{(k)}\psi_{(k)})(\sigma,\tau) \\
&\qquad \times T^{\mu\nu}(X_\alpha(\sigma,\tau))\delta^{(D)}(X_\alpha(\sigma,\tau) - X_\alpha(\sigma',\tau'))T_{\mu\nu}(X_\alpha(\sigma,\tau'))\bigg\}. \qquad (3.19)
\end{aligned}$$

The basic idea of our string solution for $QCD(U(N_c))$ is a technical improvement of Ref. [1] and consists in showing that the surface averaged propagator (3.19)

$$\langle \bar{Z}_{k\ell}[C_{X(-\pi),X(\pi)}, X_\mu(\sigma,\tau), A]\rangle_s = \mathcal{G}_{k\ell}(C_{X(-\pi),X(\pi)}, A)$$

when integrated with respect to the A-parameter as in (3.15), Now satisfies the full $U(N_c)$ non-linear Migdal-Makeenko Loop Equation (3.16) instead of the factorized Loop equations associated to the T'Hooft limit $N_c \to \infty$.

The surface average $\langle\ \rangle$, is defined by the free Bosonic action piece of (3.18) as in Sect. 2. In this context we consider $\mathcal{G}_{k\ell}(C_{X(-\pi),X(\pi)}, A)$ as the non-linear string propagator describing the "creation" of the Loop $C_{X(-\pi),X(\pi)} = C^{\text{out}}$ from the string vacuum, which is represented here by a "collapsed" point-like string initial configuration $C^{\text{in}} \equiv (x)$ (x denotes an arbitrary point of the surface which may be considered as such initial string configuration).

Let us thus, evaluate the A-derivative of $\mathcal{G}(C_{X(-\pi),X(\pi)}, A)$

$$\begin{aligned}
&\frac{\partial}{\partial A}\langle Z_{k\ell}(C_{X(-\pi),X(\pi)}, A\rangle_s \\
&\quad = \int D^F[X^\mu(\sigma,\tau)] \int D^F[\psi(k)(\sigma,\tau)] \\
&\quad \times \exp[(-S[X_\mu(\sigma,\tau), \psi_{(k)}(\sigma,\tau))] \times \psi_{(k)}(-\pi, A)\bar{\psi}_{(\ell)}(+\pi, A) \\
&\quad \times (-1) \times \frac{\partial}{\partial A}\bigg\{ \int_0^A d\tau \int_{-\pi}^{\pi} d\sigma \left[\frac{1}{2}(\partial X^\mu)^2 + \bar{\psi}_{(k)}(\gamma^a \partial_s)\psi_{(k)}\right](\sigma,\tau) \\
&\quad + \frac{\beta}{2}\int_0^a d\tau \int_{-\pi}^{\pi} d\sigma \int_0^A d\tau' \int_{-\pi}^{\pi} d\sigma' (\psi_{(k)}\bar{\psi}_{(k)}(\sigma,\tau) T^{\mu\nu}(X_\alpha(\sigma,\tau)) \\
&\quad \times \delta^{(D)}(X_\alpha(\sigma,\tau) - X_\alpha(\sigma',\tau') T_{\mu\nu}(X_\alpha(\sigma',\tau')\bigg\}. \qquad (3.20)
\end{aligned}$$

The free Bosonic term in the right-hand side of (3.20) leads to the string Laplacian as in (3.4) of Sect. 2. The free Fermion term

$$\lim_{\tau \to A^-} \bar{\psi}_{(k)}(\sigma,\tau)(\gamma^a \partial_a)\psi_{(k)}(\sigma,\tau)$$

$$\begin{aligned}
&\int D^F[X^\mu(\sigma,A)] \int D^F[\psi_k(\sigma,\tau)] \left\{ \int_{-\pi}^{\pi} d\sigma' \delta^{(D)}(X_\mu(\sigma) \quad X_\mu(\sigma') \right\} \\
&\times \frac{dX^\mu(\sigma)}{da} \cdot \frac{dX^\mu(\sigma')}{da'} \left[\sum_{p=1}^{22} \psi_k(-\pi, A) \cdot \bar{\psi}_\ell(+\pi, A) \times (\psi_p(\sigma, A) \cdot \bar{\psi}_p(\sigma, A)) \right] \\
&\times \exp(-S[X_\mu(\sigma,\tau), \psi_k(\sigma,\tau)]) \\
&\quad = \beta \int_{-\pi}^{\pi} d\sigma \int_{-\pi}^{\pi} d\sigma' \delta^{(D)}(X_\mu(\sigma) - X_\mu(\sigma') \frac{dX^\mu(\sigma)}{d\sigma} \cdot \frac{dX^\mu(\sigma')}{d\sigma'} \\
&\quad \times (\bar{Z}_{kp}[C_{X(-\pi)X(\sigma)}, X_\mu(\sigma,\tau)] \times \bar{Z}_{p\ell}[C_{X(\sigma)X(\pi)}, X_\mu(\sigma,\tau), A]), \quad (3.21)
\end{aligned}$$

vanishes as a consequence of our imposed vanished energy-momentum tensor boundary conditions on the intrinsic Fermion field. The evaluation of the boundary limit on β-term requires explicitly that the surface $\{X_\mu(\sigma,\tau)\}$ *does not possesses self-intersections of the type* $X_\mu(\sigma, A) = C_\mu^{out}(\sigma) = X_\mu(\sigma',\tau')$. *The result of this boundary limit evaluation is given explicitly by the expression below (Ref. [1]-Appendix B)*[1]

$$\sum_{p=1}^{22}\int\left(\prod_{\substack{-\pi\le\xi\le\pi\\0\le\tau\le A}} d\psi(\xi,\tau)\right)\exp\left\{\int_0^A d\tau\int_{-\pi}^{\pi} d\xi(\bar\psi_k(\gamma^a\partial_a)\psi_k)(\xi,\tau)\right\}$$
$$\times\exp\left\{-\frac{\beta}{2}\int_0^A d\tau\int_0^A d\tau'\int_{-\pi}^{\pi} d\xi\int_{-\pi}^{\pi} d\xi'(\bar\psi_k\psi_k)(\xi,\tau)T_{\mu\nu}(\xi,\tau))\right.$$
$$\left.\times\,\delta^{(D)}(X_\alpha(\xi,\tau)-X_\alpha(\xi',\tau'))T^{\mu\nu}(X(\xi',\tau'))\right\}$$
$$\times\,(\psi_k(-\pi,A)\bar\psi_\ell(\pi,A)\psi_p(\sigma,A)\bar\psi_p(\sigma,A)$$
$$=\sum_{p=1}^{22}\left[\int\left(\prod_{\substack{-\pi\le\xi\le\sigma\\0\le\tau\le A}} d\psi(\xi,\tau)\right)\right.$$
$$\times\exp\left\{\int_0^A d\tau\int_{-\pi}^{\pi} d\xi(\bar\psi_k(\gamma^a\partial_a)\psi_k)(\xi,\tau)\right\}(\psi_k(-\pi,A)\bar\psi_p(\sigma,A))$$
$$\times\exp\left\{-\frac{\beta}{2}\int_0^A d\tau\int_0^A d\tau'\int_{-\pi}^{\pi} d\xi\int_{-\pi}^{\pi} d\xi'(\bar\psi_k\psi_k)(\xi,\tau)T_{\mu\nu}(\xi,\tau))\right.$$
$$\left.\left.\times\,\delta^{(D)}(X(\xi,\tau)-X(\xi',\tau'))T^{\mu\nu}(X(\xi',\tau'))\right\}\right]$$
$$\times\left[\int\left(\prod_{\substack{\sigma<\xi<\pi\\0\le\tau\le A}} d\psi(\xi,\tau)\right)\exp\left\{\int_0^A d\tau\int_{\sigma}^{\pi} d\xi(\bar\psi_k(\gamma^a\partial_a)\psi_k)(\xi,\tau)\right\}\right.$$
$$\times\exp\left\{-\beta\int_0^A d\tau\int_0^A d\tau'\int_{\sigma}^{\pi} d\xi\int_{\sigma}^{\pi} d\xi'(\bar\psi_k\psi_k)(\xi,\tau)T_{\mu\nu}(X(\xi,\tau))\right.$$
$$\left.\left.\times\,\delta^{(D)}(X(\xi,\tau)-X(\xi',\tau'))T^{\mu\nu}(X(\xi',\tau'))\right\}\times(\psi_p(\sigma,0)\bar\psi_\ell(\pi,0))\right]\tag{3.22}$$

---

[1] Note that either crucial result below, due to the special class of surfaces chosen

$$\lim_{\substack{\varepsilon\to0\\\varepsilon>0}}\left[\int_0^A d\tau\int_0^A d\tau'\int_{-\pi}^{\sigma-\varepsilon} d\xi\int_{\sigma+\varepsilon}^{\pi} d\xi'[(\bar\psi_k\psi_k)(\xi,\tau)]T_{\mu\nu}(X(\xi,\tau))\right.$$
$$\left.\times\,\delta^{(D)}(X(\xi,\tau)-X(\xi',tau'))T^{\mu\nu}(X(\xi',\tau'))\right]=0\tag{3.21-a}$$

since our orientation tensor strings world-sheet $\{X(\xi,\tau)\}$ is such that for $\xi\neq\xi'$ and $\zeta,\zeta'\in[0,A]$

$$T^{\mu\nu}(X(\xi,\tau))\cdot T_{\mu\nu}(X(\xi',\tau'))=0\tag{21-b}$$

(these "string fixed-time" loop $X_\mu(\xi,\hat\zeta)\equiv e_\mu^{(\hat\zeta)}(\xi)$ possesses solely simple isolated self-intersections points ("eights" loops) where the non-trivial tangent lines at theses self-intersect points are supposed to be always orthogonal to each other: $T_{\mu\nu}(X(\xi,\zeta))T^{\mu\nu}(X(\xi',\zeta))=0$ as a remnant of the Fermion Exclusion Pauli Principle still acting for these Bosonic Pieces $C_\mu(\xi)$ of the full fermionic "quark trajectories").

and its unity normalization condition

$$\bar{Z}_{pp}[C_{X(\sigma)X(\sigma)}, X_\mu(\sigma,\xi), A] = 1. \tag{3.23}$$

By imposing the identification

$$g^2 N_c = \beta$$

between the $QCD(U(N_c))$ gauge coupling constant and our non linear string theory described by (3.3) we obtain the identification between the $QCD(U(N_c))$ Wilson Loop (3.17) and the surface averaged Fermion Propagator (3.19)

$$\begin{aligned}&\langle W_{k\ell}[C_{X(-\pi)X(\pi)}, A_\mu(x)]\rangle_{\text{Yang Mills } U(N_c)} \\ &\quad = \int_0^\infty dA \langle \bar{Z}_{k\ell}[C_{X(-\pi')X(\pi)}, X_\mu(\sigma,\xi), A]\rangle_s .\end{aligned} \tag{3.24}$$

The above equation is the main result of this note and generalizes to the case of $U(N_c)$ colour group our previous studies made for the t'Hooft limit of Ref. [1].

Finally we remark that by considering an ultra-violet cut-off on the space-time, $\Delta X^\mu(\sigma,\tau) \geq 1/\Lambda$, our proposed self-avoiding string theory (3.18) in the case of non dynamical $2D$-Fermions ($\langle(\psi_k\bar{\psi}_k\rangle = \mu = \text{constant}$) produces the extrinsic string with the topological invariant of string worl-sheet self-intersections number as an effective string theory for the proposed QCD string as conjectured in the first Ref. [6-8] (see Ref. [9] for this study and the enclosed Appendix).

Finally it is worth re-write (3.19) in a form where appears an interaction with an external white-noise Gaussian auxiliary anti-symetric tensor field as suggested in Ref. [1]. Namely:

$$\begin{aligned}&\exp\left\{-\frac{\beta}{2}\int_0^A d\tau \int_0^A d\tau' \int_{-\pi}^{\pi} d\tau(\bar{\psi}_{(k)}(\xi,\tau)\psi_{(k)}(\xi,\tau)) \times T^{\mu\nu}(X_\alpha(\xi,\tau))\right\} \\ &\quad \times \delta^{(D)}(X(\xi,\tau) - X(\xi',\tau'))T_{\mu\nu}(X_\beta(\xi',\tau')) \\ &= \int D^F B_{\mu\nu}(x) \exp\left\{-\frac{1}{2}\int B^2_{\mu\nu}(x)d^D x\right\} \times \exp\left\{i\int B_{\mu\nu}(x)J^{\mu\nu}(x,S)\right\}\end{aligned} \tag{3.25}$$

with the dynamical string world-sheet current

$$\begin{aligned}J^{\mu\nu}(x,S) = \beta\Big[&\int_{-\pi}^{\pi} d\xi \int_0^A d\tau T^{\mu\nu}(X(\xi,\tau))\Big(\sum_{p=1}^{22}\bar{\psi}_{(p)}\psi_{(p)}\Big)^{1/2} \\ &\times \delta^{(D)}(x - X(\xi,\tau))\Big].\end{aligned} \tag{3.26}$$

3.4 Appendix A: A Reduced Covariant String Model for the Extrinsic String

Our aim in this appendix is very modest: we write a covariant action for the elastic string and quantize in the Polyakov's path integral framework a truncated version of the covariant written theory.

Let us start our study by considering the classical action for the elastic string in the conformal gauge.

$$S_0 = \frac{1}{2\pi\alpha'}\int d^2z\rho + \gamma\int d^2z\rho\left[\left(-\frac{1}{\rho}\partial^2 X\right)^2 + i\frac{\lambda_{ab}}{\rho}(\partial_a X\partial_b X - g_{ab})\right]. \tag{A-1}$$

The string surface is described by $X = X(z)$, where X is the surface vector position in D Euclidean dimensions; z_a $(a = 1,2)$ are the coordinates of the world sheet. The first term in (A-1) is the Nambu term with the string tension equal to $1/2\pi\alpha$. The second term is the square of the extrinsic curvature with the rigidity coupling constant denoted by γ $(\gamma = \lim_{N_c\to\infty}(g^2(N_c)!)$ and $\lambda^{ab}(z)$ is a Lagrange multiplier which insures that the metric (g_{ab}) coincides with the intrinsic metrics $(\partial_a X\partial_b X)$.

Let us consider a covariant version of action (A1) by promoting $\rho(z) = g_{ab}(z)$ to be a dynamical field. This procedure yields the following action

$$\begin{aligned} S_1[X(z), g_{ab}(z), \lambda_{ab}(z)] &= \frac{1}{2\pi\alpha}\int d^2z\sqrt{g} \\ &\quad + \int d^2z\sqrt{g}\,[\gamma(-\Delta_g X)^2 + i\lambda_{ab}(g_{ab} - \partial_a X\partial_b X)]. \end{aligned} \tag{A-2}$$

Here $\sqrt{g(z)} = \det(g_{ab}(z))$ and $-\Delta g = -\frac{1}{\sqrt{g}}\partial_a(g^{ab}\partial_b)$ is the Laplace-Beltrami operator associated to the intrinsic metric $g_{ab}(z)$.

In the Polyakov's path integral quantization effective framework the partition functional for the theory (A-1) should be given by

$$Z = \int D^c[g_{ab}]D^c[X]D^c[\lambda_{ab}] \times \exp - S_1[X(z), g_{ab}(z), \lambda_{ab}(z)] \tag{A-3}$$

where the functional measures are the DeWitt covariant functional measures [1].

Let us suppose that the constraint field in approximated by the intrinsic metric $\lambda_{ab}(z) = \langle\lambda\rangle g_{ab}(z)$. (The covariant version of the usual mean field approximation $\lambda_{ab}(z) = i(\lambda)\delta_{ab}$ with (λ) a positive fixed value.) As a consequence of this hypothesis we get the truncated theory.

$$Z_{(T)} =\in D^c[g_{ab}]D^c[X]\exp - S^{(T)}[g_{ab}(z), X(z)] \tag{A-4}$$

where the truncated action theory is written as

$$S^{(T)}[g_{ab}, X] = \frac{1}{2\pi\alpha'}\int d^2z\sqrt{g(z)} + \gamma\int d^2z[(-\Delta_g X)^2] + \langle\lambda\rangle\int d^2z\sqrt{g}\,g^{ab}\partial_a X\partial_b X + \langle\lambda\rangle\int d^2z\sqrt{g(z)}. \quad \text{(A-5)}$$

For the evaluation of the X-functional integral in (A-4) we consider the non-local variable change

$$X_\mu(z) = (-i(\Delta_g)^{1/2}\vartheta_\mu)(z), \quad \mu = 1, \ldots, D.$$

Here $-i(\Delta_g)^{-1/2}$ is a well defined self-adjoint (pseudo-differential) operator. The truncated action takes the following form similar to a massive scalar field in the z domain

$$S^{(T)}[g_{ab}, \vartheta] = \left(\frac{1}{2\pi\alpha'} - \langle\lambda\rangle\right)\int d^2z\sqrt{g} + 2\langle\lambda\rangle \in d^2z\frac{1}{2}\vartheta^2 + 2\gamma\int d^2z\frac{1}{2}(\sqrt{g}\,\vartheta(-\Delta_g)\vartheta)(z). \quad \text{(A-6)}$$

The change in the (covariant) functional measure $D^c[x]$ is given by

$$D^c[x] = (\det(-\Delta_g)^{-1})^{D/2} \times D^c[\vartheta]. \quad \text{(A-7)}$$

The main step in our calculation is to define the above written function determinant as $\det^{-D/2}(-\Delta_g)$. By choosing the conformal gauge $g_{ab} = e^{\varphi}\delta_{ab}$ and evaluating the covariant Gaussian ϑ-functional intetral we obtain the partial result [1]

$$Z_{(T)} = \int D[\vartheta]\exp\left(-\frac{26-D}{48\pi}\int\left[\frac{1}{2}(\partial_a\varphi)^2 + \mu_R^2 e^{\varphi}d^2z\right]\right) \times \det{}^{-D/2}(-2\gamma\Delta_g + 2\langle\lambda\rangle) \quad \text{(A-8)}$$

where

$$\mu_R^2 = \lim_{\varepsilon\to 0}\frac{2-D}{4\pi\varepsilon} + \frac{1}{2\pi\alpha} = \langle\lambda\rangle$$

may be though as a renormalization of the bare string thension $1/2\pi\alpha'$.

We analyze now the unrenormalized functional determinant

$$\exp -S_{EFF}[\varphi] = \det{}^{-D/2}\left(-\Delta_g + \frac{\langle\lambda\rangle}{\gamma}\right).$$

By defining it by a propertime prescription we obtain the counterterms of the above written action. Explicitly

$$S_{EFF}[\varphi] = \lim_{\varepsilon\to 0} -\frac{D}{2}\int_\varepsilon^\infty \frac{dT}{T}\,Tr\left(\exp -T\left(-\Delta_g + \frac{\langle\lambda\rangle}{\gamma}\right)\right). \quad \text{(A-9)}$$

Now it is well known that the counterterms of $S_{EFF}[\varphi]$ are determined by the asymptotic expansion of the diagonal part of massive Laplace-Beltrami operator which is tabulated

$$\lim_{T\to 0} Tr\left(\exp\left(-T\left(-\Delta_g+\frac{\langle\lambda\rangle}{\gamma}\right)\right)\right)$$
$$= \int d^2z\left\{\frac{e^{\varphi}}{2\pi}\lim_{T\to 0^+}\left(\frac{1}{T}\right)-\frac{1}{2\pi}\Delta\varphi+\frac{1}{2\pi}e^{\varphi}\cdot\frac{\langle\lambda\rangle}{\gamma}\right\}(z). \quad \text{(A-10)}$$

By substituting (A-10) in to (A-9) we get straightforwardly the following counterterms associated to the two-dimensional intrinsic "mas" $\langle\lambda\rangle/\gamma$

$$\frac{D}{2}\cdot\frac{1}{2\pi}\cdot\frac{\langle\lambda\rangle}{\gamma}lg\left(\frac{1}{\varepsilon}\right)\int d^2z e^{\varphi(z)}. \quad \text{(A-11)}$$

So, on the basis of the counter term (A-11) we have the following renormalization law for the inverse of the rigidity $\beta = 1/\gamma$ (by choosing $\langle\lambda\rangle = 1$)

$$\frac{1}{\beta_R}=\frac{1}{\beta_0}-\frac{D}{2}\cdot\frac{1}{2\pi}lg(\varepsilon). \quad \text{(A-12)}$$

Equation (A-12) yields the intrinsic two-dimensional momentum dependence of the running coupling constant β

$$\beta_R(p^2)=\beta_0(p^2=0)/1-\frac{D}{2}\frac{\beta_0}{2\pi}\cdot lg\left(\frac{\varepsilon}{p^2}\right). \quad \text{(A-13)}$$

It is instructive point out the $D/2$ factor in (A-13) which appears in a natural way in our calculations.

Since it is naively expected that the string perturbative pase p^2-small ($p \in R^2$) would corresponds to the underlying QCD field theory at its non-perturbative phase $k^2 \to +\infty$ ($k \in R^4$), one can see that (A13) suggests a natural explanation from the QCD's String Representation for the "strange" QCD field theory description of the asymptotic behavior for the coupling constant at large N_c, namely

$$\lim_{k^2\to\infty}\left(\lim_{N_c\to\infty}(g^2N_c)_{\text{ren}}(k^2)\right)=0. \quad \text{(A-14)}$$

As a general conclusion, one can see that still exists a great deal of not completely understood phenomena in QCD out of non-analytical field theoretic continuum approaches-lattice approximations.

3.5 Appendix B: The Loop Space Program i the Bosonic $\lambda\phi^4 - O(N)$-Field Theory and the QCD Triviality for R^D. $D > 4$

Let us start our study by considering the (bare) generating functional of the Green's functions of the $O(N)$ (symmetric phase) $\lambda\phi^4$ field theory in a D-dimensional Euclidean space-time

$$Z[J^a(x)] = \int \prod_{a=1}^{N} d\mu[\Phi^a(x)] \times \exp\Big\{ -\frac{\lambda_0}{4} \int d^D x \left(\sum_{a=1}^{N} \Phi^a(x)^2 \right)^2 (x) - \int d^D x \left(\sum_{a=1}^{N} J^a(x)\Phi^a(x) \right) \Big\} \tag{B-1}$$

where $\Phi^a(x)$ denotes a N-component real scalar $O(N)$ field, (μ_0, λ_0) the (bare) mass and coupling parameters and the Gaussian functional measure in (B-1) is

$$\prod_{a=1}^{N} d\mu[\Phi^a(x)] = \prod_{a=1}^{N} \left(\prod_{x\in R^D} d\Phi^a(x) \right) \exp\Big\{ -\frac{1}{2} \int d^D x \left(\sum_{a=1}^{N} (\partial_\mu \Phi^a) \right)^2 (x) + \mu_0^2 \sum_{a=1}^{N} (\Phi^a)^2(x) \Big\}. \tag{B-2}$$

Now, in order to get an effective expression for the functional integrand (B-2), where we can evaluate the Φ^a functional integrations, we write the intersection $\lambda\phi^4$ term in the following form

$$\exp\left\{ -\frac{\lambda_0}{4} \int d^D x \left(\sum_{a=1}^{N} (\Phi^a(x))^2 \right)^2 \right\} = \int d\mu[\sigma] \cdot \exp\left\{ -i \int d^D x \sigma(x) \left(\sum_{a=1}^{N} (\phi^a(x))^2 \right) \right\} \tag{B-3}$$

where $\sigma(x)$ is an ausiliary scalar field and the σ functional measure in (B-3) is given by

$$d\mu[\sigma] = \left(\prod_{x\subset R^D} d\sigma(x) \right) \exp\left\{ -\frac{1}{2} \int d^D x \frac{2}{\lambda_0} \sigma^2(x) \right\} \tag{B-4}$$

with covariance

$$\langle \sigma(x_1)\sigma(x_2) \rangle_\sigma = \int d\mu[\sigma]\sigma(x_1)\sigma(x_2) = \frac{\lambda_0}{2} \delta^{(D)}(x_1 - x_2). \tag{B-5}$$

The last result allows us to consider the $\delta(x)$ field as a random Gaussian potential with noise's strength $\frac{\lambda_0}{2}$.

After substitution of (B-4) into (B-2), we can evaluate explicitly the Φ-functional integrations since they are of Gaussian type. We, thism get the result

$$Z[J^a(x)] = \int d\mu[\sigma] \det{}^{-N/2}(-\Delta + \mu_0^2 - 2i\sigma)$$
$$\times \exp\left\{\frac{1}{2}\int d^Dx d^Dy J^a(x)(-\Delta + \mu_0^2 - 2i\sigma)\delta_{ab}J^B(y)\right\}. \quad \text{(B-6)}$$

At this point of our study we implement the main idea: by following symanzik's analysis, we express the σ-functionals integrands in (B-6) as functional defined in the Feynman-Kac-Wiener space of Random paths by making use of the well known random path respesentation for the non-realativistic euclidean propagator of a particle of mass μ_0 in the presence of the external random Gaussian potential $\sigma(x)$

$$(-\Delta + \mu_0^2 - 2i\sigma)^{-1}(x,y) = \int_0^\infty d\zeta G(x,y,\sigma)(\zeta) \quad \text{(B-7)}$$

$$\log\det(-\Delta + \mu_0^2 - 2i\sigma) = -\int_0^\infty \frac{d\zeta}{\zeta}\int d^Dx F(x,x,\sigma)(\zeta) \quad \text{(B-8)}$$

where the non-relativistic propagator is given by

$$G(x,y,\sigma)(\zeta) = \int d\mu\{w_{xy}^{(\zeta)}\}e^{i\int d^D\sigma(z)j(w_{wy}^{(\zeta)}(z))} \quad \text{(B-9)}$$

with the Feynman-Kac-Wiener path measure

$$d\mu[w_{xy}^{(\zeta)}] = \left(\prod_{\substack{0<\alpha<\zeta \\ w(0)=x \\ w(0)=y}} dw[\alpha]\right)\exp\left\{-\frac{1}{2}\int_0^\zeta \left(\frac{dw}{d\alpha}\right)^2 - \frac{1}{2}\mu_0^2\zeta\right\} \quad \text{(B-10)}$$

and the (random) world-line currents defined by

$$j(w_{xu}^{(\zeta)})(z) = \int_0^\zeta \delta^D(z - w_{xu}^{(\zeta)}(\alpha))\,d\alpha. \quad \text{(B-11)}$$

So, we obtain the proposed reformulation of $\lambda\phi^4 O(N)$-theory as a theory of random paths $\{w_{xy}^{(\zeta)}(\alpha)\}$ in the presence of a random Gaussian potential

$$\begin{aligned}Z[J^a(x)] = \int d\mu[\sigma] \times \exp\Big\{\frac{N}{2}\int_0^\infty \frac{d\zeta}{\zeta}\int d^D x \int d\mu[w_{xx}^{(\zeta)}] \\ + \exp\left(i\int d^D z\sigma(z) j(w_{xx}^{(\zeta)})(z)\right)\Big\} \\ \times \exp\Big\{\frac{1}{2}\int d^D x^D y \sum_{s=1}^{N} j_a(x)\Big[\int_0^\infty d\zeta \int d\zeta[w_{xy}^{(\zeta)}] \\ \times \exp\left(i\int d^D z\sigma(z) j(w_{wy}^{(\zeta)}\right)\Big]\delta_{ab}\cdot J_b(y)\Big\}. \qquad \text{(B-12)}\end{aligned}$$

We shall use ther andom path formulation (B-12) to analysie the correlation functions of the $\lambda\phi^4$ theory. As a useful remark, we note by using (B-12) that the general k-point (bare) correlation function possesses the general structure for free fields

$$\langle \Phi_{i_l}(x_1)\dots\Phi_{i_k}(x_k)\rangle_\Phi = \begin{cases} 0 & \text{if } k = 2j+1 \\ \displaystyle\sum_{\substack{(2j+1)\\ \ell\text{-pairings}}} \langle\Phi_{i_l}(x_{\ell_1})\Phi_{i_2}(x_{\ell_2})\rangle_\Phi \dots \langle \Phi_{i_{k-1}}(x_{\ell_{2j-1}})\Phi_{i_k}(x_{\ell_{2j}})\rangle_\Phi & \text{if } k = 2j \end{cases} \qquad \text{(B-13)}$$

where the quantum averages $\langle\ \rangle_\Psi$ in (B-13) are defined by the $\lambda\phi^4$ partition functional $Z[0]$ (see (B-1) with $J^a(x)\equiv 0$).

Because of this result, we have solely to study the properties of the 2-point correlation function

$$\begin{aligned}\langle\Phi_{i_l}(x_1)\Phi_{i_2}(x_2)\rangle_\Phi = \delta_{i_1 i_2}\Big\langle \int_0^\infty d\zeta d\mu[w_{x_1 x_1}^{(\zeta)}] \\ \times \exp\Big\{i\int d^D z\sigma(z)\cdot j(w_{x_1 x_2}^{(\zeta)})(z)\Big\} \\ \times \exp\Big\{\frac{N}{2}\int_0^\infty \frac{d\zeta}{\zeta}\int d^D x\int d\mu[x_{xx}^{(\zeta)}] \\ \times \exp\Big[i\int d^D z\sigma(z) j(x_{xx}^{(\zeta)}(z))\Big]\Big\}\Big\rangle_0. \qquad \text{(B-14)}\end{aligned}$$

Let us evaluate the σ-functional averages $\langle\ \rangle_\sigma$ in (B-4) and (B-5)). For this task we expand the "close path term"in powers of N. Explicitly

$$\langle \Phi_{i_1}(x)\Phi_{i_2}(y)\rangle = \delta_{i_1 i_2} \sum_{k=0}^{\infty} \left(\frac{N}{2}\right) \Big\{ \prod_{\ell=1}^{k} \int_0^\infty \frac{d\zeta_\ell}{\zeta_\ell} \int d^D x_\ell$$
$$\times \int d\mu[w^{(\zeta)}_{x_\ell x_\ell}] \times \int_0^\infty d\zeta \times \int d\mu[w^{(\zeta)}_{xy}] \Big\}$$
$$\times \Big\langle \exp\Big\{ i \sum_{\ell=1} k \int d^D z_\ell \sigma(z_\ell) \times j(w^{(\zeta_\ell)}_{x_\ell x_\ell})(z_\ell)$$
$$+ i \int d^D \sigma(z) j(w^{(\zeta)}_{xy})(z) \Big\} \Big\rangle \tag{B-15}$$

and since the σ-average in (B-15) is of the Gaussian type we can perform it exactly. The result reads

$$\langle \Phi_{i_1}(x)\Phi_{i_2}(y)) = \delta_{i_1 i_2} \sum_{k=0}^{\infty} \left(\frac{N}{2}\right) \Big\{ \prod_{\ell=1}^{k} \int_0^\infty \frac{d\zeta_\ell}{\zeta_\ell} \int d^D x_\ell \int d^D x_\ell \int d\mu[w^{(\zeta_\ell)}_{x_\ell x_\ell}]$$
$$\times \int_0^\infty d\zeta \times \int d\mu[w^{(\zeta)}_{xy}] \times \exp\Big\{ -\frac{\lambda}{4} \Big[\Big(2 \times \sum_{\ell \neq \ell'}^{k} \int_0^{\zeta_\ell} d\alpha_\ell \times \int_0^{\zeta_{\ell'}} d\alpha_{\ell'}$$
$$\times \delta^{(D)}(w^{(\zeta_\ell)}_{x_\ell x_\ell} \alpha_\ell) - w^{(\zeta_{\ell'})}_{x_{\ell'} x_{\ell'}}(\alpha_{\ell'}))\Big)$$
$$+ \Big(\sum_{\ell=\ell'}^{k} \int_0^{\zeta_\ell} d\alpha_\ell \int_0^{\zeta_\ell} \delta^{(D)}(w^{(\zeta_\ell)}_{x_\ell x_\ell}(\alpha_\ell) - w^{(\zeta_\ell)}_{x_\ell x_\ell}(\alpha_{\ell'}))\Big)$$
$$+ \Big(2 \times \sum_{\ell=1}^{k} \int_0^{\zeta_\ell} d\alpha_\ell \times \int_0^{\zeta} d\alpha \delta^{(D)}(w^{(\zeta_\ell)}_{x_\ell x_\ell}(\alpha_\ell) - w^{(\zeta)}_{xy}(\alpha) - w^{(\zeta)}_{xy}(\alpha))\Big)$$
$$+ \Big(\int_0^{\zeta} d\alpha \int_0^{\zeta} d\alpha' \delta^{(D)}(w^{(\zeta)}_{xy}(\alpha) - w^{(\zeta)}_{xy}(\alpha 1))\Big) \Big] \Big\}. \tag{B-16}$$

The above expression is the two-point correlation function of the $\lambda\Phi^4 - O(N)$-theory expressed as a sustem of interacting random paths with a repulsive self-interaction at these points where they crosses themselves.

Now we can offer a topological explanation for the theory triviality phenomenon for $D > 4$. At first, we note that the correlation function (B-17) will differ from the free one, namely

$$\langle \Phi_{i_1}(x)\Phi_{i_2}(y)\rangle_{\mathrm{FREE}} = \delta_{i_1 i_2} \left(\int_0^\infty d\zeta d\mu[w^{(\zeta)}_{xy}] \right) \tag{B-17}$$

if the path intersections implied by the delta functions in (B-16) are non-empty sets in the R^D space-time. We intend to argument that those intersection sets are empty for space-time with dimensionality greater than four. At first we recall some well-known concerpts of topology the topological Hausdorff dimension of a set A embedded in R^D is d (with d being a real number) if the inimum number of D-dimensional spheres of radius γ needed to cover it, grow like γ^{-d} when $r \to 0$. The rule for (generical) intersections for sets A and B (both are embeddeds in R^D) is given by

$$d(A \cap B) = d(A) + d(B) - D \tag{B-18}$$

where a negative Hausdorff dimension means no (generical) intersection or equivalently the set $A \cap B$ is empty.

As is well known the Hausdorff dimension of the random paths in (B-16) is 2. A direct application of the rule (B-18) gives us that the intersection sets in (B-16) possesses a Hausdorff dimension $4 - D$. So, for $D > 4$ these sets are empty and leading to the triviality phenomenon (see (B-17)).

Finally we make some comments on the analyses of the divergencies in the random path expression (B-16) for $D \leq 4$. As a first observation we note that all the path integrals involved in (B-16) can be exactly evaluated by making a power series in λ_0. The resulting proper-times ζ integrals will in general be divergents. Bu using a regularization (such as a cut off for small proper-times) one can show that the divergencies can be absorved by a renormalization of the bare mass μ_0 and the action path term in (B-16) (or equivalently, a wave-function and λ_0-coupling renormalization in the field formulation (B-1)).

At this point of our remarks and comments, it is worth to point out that there is no simple relation between our random loop space approach for QCD where the loop defining the string world-sheet boundary is anon-differontiable path and representing rigorously the functional determinant associated to the matter ontent E [this means that there is no pure Yang-Mills quantum theory without matters source in our approach (no rings of Gluons!)]; and others approach based on suibable3 supersymmetric σ-models formulations for conformal superstrings moving in non quantum back-grounds (see [10] and [11]). Note that in this case there is still no true derivation of this string/gauge field duality from first principles.

In our string representation for Bosonic QCD as we have proposed in this note, one can see that the Hausforff dimension of the continuous manifolds sampled by the (euclidean) quantum string vector position is four (a very rough Brownion Bosonic Surface filling u any four-volume in R^4). However, it is expected that the Hausdorff dimension of the manifold sampled by the $2D$-Fermion Field should be minus two. Combining these results one can see that the effective Hausdorff dimension of the QCD string world-sheet is two, so allowing one assplies all concepts of classical smooth Differential Topology and Geometry. Ilf all these results turn out to be rigorous, one can see that our self-avoiding fermionic string representation gives a "proof" that $QCD(U(N_c))$ should be expected to be a trivial quantum field theory (with on infrared cut off!) for space-times dimension greater than four.

Finally, we should remark that our proposal for string representations in QCD has no apparent overlap with those proposal relying heavily in the existence of the string Liouville field theory as a bonafide $2D$ Field Theory as proposed in [12] even if they can be interpreted as an extra (umphysical) five dimension coordinate after some conformal impositions in the non-critical string theory.

It appears interesting to remark that these Kaluza-Klein string representations for $N = 4$ Supersymmetric QCD may be considered as "modern/geometrical/topological" version of the old beautiful result in String Theory that Strings with $U(N)$ Chan-Paton factors leads formally to Massless and Massive Yang-Mills scattering amplitude in its low energy limit of vanishing Regge Sloppe limit [13].

3.6 References

[1] Botelho, L.C.L.: Methods of Bosonic and Fermionic Path Integral, Representations – Continuum Random Geometry in Quantum Field Theory. Nova Science, New York (2008).

[2] Botelho, L.C.L.: Phys. Rev. D **40**, 660 (1989).

[3] Botelho, L.C.L.: Phys Rev. D **41**, 3283 (1990).

[4] Botelho, L.C.L.: Phys. Lett. B **152**, 358 (1985).

[5] Botelho, L.C.L.: Rev. Bras. Fis. **16**, 279 (1986).

[6] Polyakov, A.: Nucl. Phys. B 268, 406 (1986).

[7] Peliti, L., Leibler, S.: Phys. Rev. Lett. **54**, 1690 (1980).

[8] Duplanter, B.: Commun. Math. Phys. **85**, 221 (1982).

[9] Karanikas, A.I., Ktorides, C.H.: Phys. Lett. B **235**, 90 (1990).

[10] Fradkin, E.S., Tseytlin, A.A.: PLB ISS, 316 (1981).

[11] Maldacena, M.: Phys. Rev. Lett. **80**, 4859 (1988).

[12] Gibson, S.S.: Klebanov, I.R., Plyakov, A.M.: Gauge theory correlators from non-critical string theory. arXiv:hep-th/19802109.

[13] Botelho, L.C.L.: PRD 35(4), 1515–1518 (1987).

Chapter 4

The Formalism of String Functional Integrals for the Evaluation of the Interquark Potential and Non Critical Strings Scattering Amplitudes[1]

We present news path integral studies on the Polyakov Non-Critical and Nambu-Goto critical strings theories.

We also evaluate the long distance asymptotic behavior of the interquark potential on the Nambu-Goto string theory with an extrinsic term. We also propose an alternative and new view to covariant Polyakov's string path integral with a fourth-order two-dimensional quantum gravity.

4.1 Introduction

One of the most promising mathematical formalism for a physically sensible description of strong interactions is quantum chromodynamics. In strong interaction physics the image of an physically detectable mesonic quantum excitation is, for instance, the quantum mechanical color invariant probability of the appearance of a pair quark-antiquark bounding a space-time non abelian gluon surface connecting both the pair's particle.

It appears tantalizing for mathematical formulations to consider as fundamental gauge invariant dynamical variable, the famous quantum Wilson loop, with the loop C (defining the non-abelian holonomy factor); being given by the quark-antiquark (space-time) Feynman trajectory ([1], [2]) (see also chapters 1 and 2).

It is thus searched loop space dynamical equations (at least on the Lattice on the formal grounds) for the quantum Wilson loop wich supports hopes for a complete string solution for Q.C.D., at least at the large number of colors ([1]) (see also chapter 3).

[1]Complements to Chapter 2.

It is natural thus to study path integral geometrodynamical string propagators in order to make such connection between string and QCD mathematically more precise ([3]) (also chapter 2 and 3).

Chapter 4 is organized as follows: In section 2, we present a detailed stuty of the Nambu-Goto string path integral. In section 3 and appendixes A and B, we extend the results of section 2 to the case of the presence of extrinsic geometry. In the section 3 also, we present a new result on the large distance asymptotic behavior of the interquark potential by supposing QCD represented by the Nambu-Goto extrinsic string at large distance.

In section 4 and appendix C, we present a new proposal for the Polyakov's Non-Critical String.

4.2 Basics Results on the Classical Bosonic Surface Theory and the Nambu-Goto String Path Integral

Let us start our considerations by considering a given continuously differentiable globally orientable; compact surface S, with a boundary given by a non self-intersecting smooth curve C and fully immerse on the space-time R^D. Its mathematical description is described by a $C^1(\Omega)$ two-dimensional vector field $X_\mu(\xi_1, \xi_2)$, $(\mu = 1, \ldots, D)$ on a two-dimensional domain Ω (compatible with the fixed-prescribed topology of S).

The Nambu-Goto-Buff-Lovett-Stillinger ([1]) area is given by the usual geometric integral $((\xi_1, \xi_2) := \xi)$

$$A(S_c) = \int_\Omega (\det\{h_{ab}(\xi)\})^{1/2} d^2\xi \tag{4.1}$$

with $h_{ab}(\xi) = (\partial_a X^\mu \partial_b X_\mu)(\xi)$ denotes the metric tensor induced on the surface S_c.

Note that the scalar of curvature

$$R(X^\mu) = \{h^{ab}(Ric)_{ab}\}(X^\mu(\xi));$$

$$\begin{aligned}\Big[Ric_{ab}(X^\mu(\xi)) = R_{abcd} &= \frac{1}{2}\Big(\frac{\partial^2 h_{ac}}{\partial x^b \partial x^d} - \frac{\partial h_{bc}}{\partial x^a \partial x^d} - \frac{\partial^2}{\partial x^b \partial x^c} \\ &+ \frac{\partial^2 g_{cd}}{\partial x^a \partial x^c} + g_{rs}(\Gamma^r_{ac}\Gamma^s_{bd} - \Gamma^r_{ad}\Gamma^s_{bc})\Big) \quad R^c_{acb}(X^\mu(\xi)) \\ &= (\partial_c \Gamma^c_{ba} - \partial_b \Gamma^c_{ca} + \Gamma^d_{ba}\Gamma^c_{cd} - \Gamma^d_{ca}\Gamma^c_{bd})(X^\mu(\xi))\end{aligned}$$

with naturally

$$\Gamma^a_{bc} = \left(\frac{1}{2} h^{ar}\left(\frac{\partial}{\partial \xi^b} h_{rc} + \frac{\partial}{\partial \xi^c} h_{rb} - \frac{\partial}{\partial \xi^r} h_{bc}\right)\right)\Big]$$

must satisfy the topological constraint that originates from the Gauss theorem. For the case of boundaryless compact surfaces without "handles"

$$\frac{1}{2\pi}\left(\int_{\Omega}(\sqrt{h}\,R)(X^{\mu}(\xi))d^{2}\xi\right)=2-2g, \tag{4.2}$$

where g denotes the number of "holes" of S_c.

Note again that the domain Ω must be compatible with the surface topology ([4]).

The most important property of the above written functionals on the Riemman surface S_c {(we point out that under the above mathematical imposed conditions on the surface S_c added with an additional assumption of C^{∞}-differentiability, it is a mathematical consequence that S_c may be endowed now with a complex structure, turning it a Riemman surface, paramtrized by holomorphic vector fields on $\Omega \subset R^2$; $X^{\mu}(\xi_1, \xi_2) = X^{\mu}(z)$ with $z = \xi_1 + i\xi_2$, $i = \sqrt{-1}$)} is the local invariance under the group of local diffeomorphism of the surface S

$$\begin{aligned} \xi'_a - \xi_a &= \delta\xi_a =: E_a(\xi) \\ \delta X_{\mu}(\xi) &= \varepsilon_a(\xi)\partial^a X_{\mu}(\xi). \end{aligned} \tag{4.3}$$

It is worth that the above pointed out invariance under the diffeomorphism local group of the surface S_c, can be extended to the global diffeomorphism case, only for trivial topological surfaces with $g = 0$ (no "handles") (see eq. (4.2)).

Let us thus follow R.P. Feynman in his theory of path integration sum over "classical-random" histories of a quantum system (with a classical mechanical system countor part as in our case: The Correspondence Principle of Quantum Mechanics in action), in order to quantize our string theory (the curve C can be considered as our classical string and the surface S_c denotes its euclidean quantum trajectory in the space-time R^D), through an anihillation string process).

$$\begin{aligned} G(C) = \sum_{g=0}^{\infty}\Bigg\{\int_{X_{\mu}(\xi)|_{\xi\in\partial\Omega}=C^{\mu}(\sigma)} & d_h\mu[X_{\mu}(\xi)] \times \exp\left\{-\frac{1}{2\pi\alpha'}A(S_c)\right\} \\ \times\delta^{(F)}\left(\left[\int_C K_h(\sigma)d\sigma + \frac{1}{2\pi}\int(\sqrt{h}R)(X^{\mu}(\xi)) - (2-2g)\right]\right)\Bigg\}. \end{aligned} \tag{4.4}$$

Here α' denotes the Regge slope parameter which has the dimension of inverse of mass square (in universal units $\hbar = c = 1$). Here $K_h(\sigma)$ is the geodsica curvature of the surfaces boundary C.

The above written non-polinomial $2D$ quantum field theory is somewhat complicated in its methematical perturbative calculational structure, since $d\mu_h(X^\mu(\xi))$ is not the usual Feynman product measure, but the "weighted" product Feynman measure ([chapter 2]), in order to preserve the invariance of the geometrodynamical string propagator eq. (4.4) under the action of the main theory's symmetry eq. (4.3). Explicitly

$$d_h\mu[X^\mu(\xi)] := \prod_{\xi\in\Omega}[(h(X(\xi)))^{1/4}\, dX_\mu(\xi)]. \tag{4.5}$$

The above written local diffeomorphism invariant functional measure on the string vector position is obtained form the local diffeomorphism invariant functional Riemann metric

$$||\delta X_\mu||^2 = \int_\Omega (h(X^\mu(\xi))^{1/2}(\delta X_\mu)(\xi)(\delta X_\mu)(\xi)d^2\xi. \tag{4.6}$$

However the first step to evaluate the so called Nambu-Goto string path integral eq. (4.4) is to consider the quantization process as "quantum fluctuations" around the classical system motion (R.P. Feynman):

$$X_\mu(\xi) = X_\mu^{CL}(\xi) + \left(\sqrt{\pi\alpha'}\right)\overline{X}_\mu(\xi) \quad (\hbar = 1). \tag{4.7}$$

The classical dynamics is given by Euoer-Lagrange equations associated to the surface area functional under the topological constraint eq. (4.2). And before proceeding, we remark that is only in this step and on the topological form of Ω where the topological constraint is taken into account. As a consequence we can disregard explicitly the functional topological constraint on eq. (4.4). We have thus, the classical motion equations (a Dirichlet nonlinear problem)

$$\begin{aligned} &\Delta_h X_\mu^{CL}(\xi) = 0 \\ &X_\mu^{CL}(\xi)|_{\xi\in\partial\Omega} = C^\mu. \end{aligned} \tag{4.8}$$

Here Δ_h denotes the second-order elliptic operator called Laplace-Beltrami associated to the metric

$$h_{ab}(\xi) = h_{ab}(X_\mu(\xi)) = (\partial_a X^\mu \partial_b X_\mu)(\xi).$$

$$\Delta_h = \left(\frac{1}{\sqrt{h}}\partial_a(\sqrt{h}\, h^{ab}\partial_b)\right)(\xi). \tag{4.9}$$

At this point, one most use formally, at least for trivial topological surfaces S_c, the hypothesis of the global extension of the local diffeomorphism group of the surface S_c, in order to use globally on S_c the conformal coordinate system

$$\xi_1' = \xi_1'(\xi_1, \xi_2); \quad \xi_2' = \xi_2'(\xi_1, \xi_2).$$

$$h_{ab}(X'^{\mu}(\xi')) = \rho(X'^{\mu}(\xi'))\delta_{ab}. \tag{4.10}$$

The new coordinates are given by the global (or local) diffeomorphic solutions of the Laplace-Beltrami: equations given below ($i = \sqrt{-1}$)

$$d\xi_1' + id\xi_2' = \lambda(\xi_1, \xi_2)\Big(\sqrt{h_{11}(X_\mu(\xi))}\, d\xi_1 + \frac{(h_{12}(X_\mu(\xi)) + i\sqrt{h(X_\mu(\xi))})}{\sqrt{h_{11}(X^\mu(\xi))}}\, d\xi_2\Big) \tag{4.11-a}$$

$$d\xi_1' - id\xi_2' = \mu(\xi_1, \xi_2)\Big(\sqrt{h_{11}(X_\mu(\xi))}\, d\xi_1 + \Big(\frac{(h_{12}(X_\mu(\xi)) - i\sqrt{h(X_\mu(\xi))})}{\sqrt{h_{11}(X^\mu(\xi))}}\Big) d\xi_2\Big). \tag{4.11-b}$$

In this new coordinate system on S_c ([4])

$$h_{ab}(X_\mu'(\xi')) = \rho^2(\xi')\delta_{ab} = \left(\frac{1}{|\lambda\mu|(\xi, \xi')}\right)\delta_{ab}. \tag{4.12}$$

If one now choose the conformal gauge for the surface S_c $h_{ab}(\xi) = e^{\varphi(\xi)}\delta_{ab}(\varphi(\xi) = 2\ln\rho(\xi))$, one reduces the string non Linear elliptic problem eq. (4.8) to the well-studied Dirichlet problem in Ω

$$\begin{cases} \Delta_{h=\delta_{ab}} X_\mu'(\xi') = 0 \\ X_\mu'(\xi')|_{\xi' \in \partial\Omega'} = C'^{\mu}. \end{cases} \tag{4.13}$$

Note that the problem full solution is given by

$$X_\mu(\xi) = X_\mu'(\xi'(\xi)). \tag{4.14}$$

The solution of eq. (4.13) can be always be analyzed by methods of conformal complex variable methods ([4], [5]) specially for the trivial topological case (connected planar Ω') ([4]).

Unfortunatelly, it appears that the resulting quantum theory (path integral) for eq. (4.4)–eq. (4.5) still remains as an open problem, from a purely perturbative approach around the loop-expansions in an α' – power expansion.

However in the simply case of the domain $\Omega_{(R,T)}$ being a rectangle of sizes $0 \le \xi_1 \le R$; $0 \le \xi_2 \le T$, an exact one-loop result can be obtained (Note that $C = \partial\Omega$).

It is straightforward to see that

$$S[\overline{X}^{\mu}(\xi_1,\xi_2)] = \frac{1}{2\pi\alpha'}(RT) + \frac{1}{2}\int_0^T d\xi_2 \int_0^R d\xi_1 (\partial\overline{X}^{\mu}(\xi))^2 + O(\alpha'^2). \tag{4.15-a}$$

$$d_h[\overline{X}^{\mu}(\xi_1,\xi_2)] = D^F[\overline{X}^{\mu}(\xi_1,\xi_2)] + O(\alpha'). \tag{4.15-b}$$

This leads to the result

$$G(C_{[R,T]}) = \exp\left\{-\frac{1}{2\pi\alpha'}RT\right\} \times \left(\det_{\Omega_{(R,T)}}(-\Delta)\right)^{-(\frac{(D-2)}{2})}. \tag{4.16}$$

By noting the explicit (non-trivial) evaluation of the functional determinant of the Laplacean on the Torus $\Omega_{(R,T)}$ with Dirichlet conditions ([7], vol II):

$$\left(\det_{\Omega_{(R,T)}}(-\Delta)\right)^{-1/2} = \frac{1}{(\frac{T}{R})^{1/2}}\left[\left(e^{-\frac{4\pi T}{R}}\right)^{-1/24} \times \left(\prod_{n=1}^{\infty}(1-e^{-\frac{2\pi n T}{R}})^{-2}\right)\right], \tag{4.17}$$

one can see that the quantum strong ground state has the "confining behavior" with the Coulomb-Lscher term as its energy on this one-loop approximation

$$E_{\text{Vacuum}}(R) = \lim_{T\to+\infty}\left\{-\frac{1}{T}\ln G(C_{(R,T)})\right\} = \left(\frac{R}{2\pi\alpha'}\right) - \left(\frac{\pi(D-2)}{6}\cdot\frac{1}{R}\right). \tag{4.18}$$

Unfortunatelly ths string scattering amplitudes were never evaluated in a undisputable form in this Nambu-Goto strong theory, unless on the light-cone gauge by S. Mandelstam ([3], [6]).

As a consequence of the above mathematical aspects on the Nambu-Goto quantum string theory, A.M. Polyakov has proposed a new functional integral approach to overcame some of the above difficulties.

The complete mathematical exposition of the A.M. Polyakov propose will be exposed (in details) on next section (see also chapter 2).

As a final comment let us try to evaluate the Nambu-Goto string path integral through the use of the conformal gauge as given by eq. (4.12). In this case (the so called light – cone gauge), we have the following constraints, after introducing the complex-light-cone euclidean coordinate on the domain Ω

$$\begin{aligned} \partial_+ &= \frac{(\partial_{\xi_1'} - i\partial_{\xi_2'})}{2} = \partial_z \\ \partial_- &= \frac{(\partial_{\xi_1'} + i\partial_{\xi_2'})}{2} = \partial_{\bar{z}}. \end{aligned} \tag{4.19}$$

We have that

$$\partial_+ X'_\mu \partial_+ X'_\mu = \partial_- X'_\mu \partial_- X'_\mu (\Leftrightarrow \partial_{\xi_1'} X'_\mu \partial_{\xi_2'} X'_\mu = h_{12} = h_{21} = 0) \tag{4.20-a}$$

$$\sqrt{h(X'_\mu(\xi'))} = \partial_+ X'_\mu \partial_- X'_\mu (\Leftrightarrow h_{11} = h_{22} = \partial_+ X'_\mu \partial_- X'_\mu). \tag{4.20-b}$$

In this gauge the path integral eq. (4.4) for $C = \{\phi\}$ (the string partition functional) takes the form (for the simple case of trivial topology surface $g = 0$).

$$\begin{aligned} Z = \int d_h\mu[X'_\mu(z)] \exp\Big\{ &- \frac{1}{2\pi\alpha'} \int_\Omega \left(\frac{dz \wedge d\bar{z}}{2i}\right) \\ &\times (\partial_z X'^\mu)(\partial_{\bar{z}} X'^\mu)(z,\bar{z})\Big\} \\ &\times \delta^{(F)}[(\partial_z X'_\mu \partial_z X'_\mu)] \times \delta^{(F)}[(\partial_{\bar{z}} X'_\mu \partial_{\bar{z}} X'_\mu)]. \end{aligned} \tag{4.21}$$

Note that in the practical use of eq. (4.1) to evaluate string observable average, one already uses the observable on the light-cone gauge eq. (4.20), which by its turn suppress the explicitly use of the above written delta functionals insuring that Feynman-Wiener measure $d_h\mu[X'_\mu(\xi)]$ is already in this gauge.

It has been proved by my self ([1]) that the non-linear measure $d_h\mu[X'_\mu(\xi)]$ on the light-cone gauge can be related by the A.M. Polyakov-Stromingh-Fugikawa conformal anomaly factor to the Feynman-Wiener simply weighted measure

$$\begin{aligned} d_h\mu[X'_\mu(\xi)] &= \left(\prod_{(\mu,z,\bar{z})} dX_\mu(z,z')\right) \\ &\times \left\{\frac{\delta^{(e)}(0)}{4} \int_\Omega \ln(h(X_\mu(z,z)))dz\, d\bar{z}\right\} \\ &\times \exp\left\{-\frac{(26-D)}{48\pi} \int_\Omega \frac{dz d\bar{z}}{2} \left[\frac{\partial_+(\partial_+ X_\mu \partial_- X_\mu) \cdot \partial_-(\partial_+ X_\mu \partial_- X_\mu)}{(\partial_+ X_\mu \partial_- X_\mu)^2}\right](z,z)\right\}. \end{aligned} \tag{4.22}$$

One can see thus that only at $D = 26$, the pure bosonic Nambu-Goto string is described by Marsless scalar fields on the string domain Ω, if One uses formally the Bollini-Giambiagi dimensional regularization scheme to assign the unity value for the tad-poles Feynman diagramms[1] ([1], [4]), through a not completely understood Feynman diaggrammatics for two-dimensional (mathematically ill defined) Massless scalar fields for $\Omega = R^2$ (the so called Coleman theorem ([1])).

However, One can follows the Virassoro-Sakita proposal to evaluate string scattering amplitudes using scalar vertex without bothering ourselve with gauge fixed technical details ([3]).

We thus use the path integral eq. (4.21) for $\Omega = R^2$ in order to evaluate the closed string (scalar) N-point scattering amplitude at $D = 26$ (with $\delta^{(2)}(0) = 0$)

$$A(\rho_1^\mu, \dots, \rho_N^\mu) = \frac{1}{Z}\left\{ \int D^F[X_\mu(\xi)] \exp\left[-\frac{1}{\pi\alpha'} \int_{R^2} (\partial X_\mu \cdot \partial X_\mu)(\xi) d^2\xi \right] \right.$$
$$\left. \times \left[\int_{R^{2N}} d^2\xi_j \prod_{j=1}^{N} \exp(i\rho_\mu^j \cdot X_\mu)(\xi_j) \right] \right\}. \tag{4.23}$$

Since eq. (4.23) is formally a Gaussian Functional Integer and we can re-write the scalar vertexs as string vector position source ($i = \sqrt{-1}$)

$$\left(\prod_{j=1}^{N} \exp(i\rho_\mu^j X_\mu)(\xi_j) \right)$$
$$= \exp\left\{ i \left[\sum_{j=1}^{N} \int_{\mathbb{R}^2} (\rho_\mu^j \delta^{(2)}(\xi - \xi_j)) X_\mu(\xi) d^2\xi \right] \right\}, \tag{4.24}$$

and by using the dimensional regularization technique to vanish the tad-pole term

$$\exp\left\{ -\sum_{j=1}^{N} (\rho_\mu^j)^2 (-\Delta)^{-1}(\xi_j, \xi_j) \right\} = 1. \tag{4.25}$$

One gets the Veneziana N-point amplitude as a result in R^{26} ([3]. [6])

$$A(\rho_1^\mu, \dots, \rho_N^\mu) = \int_{R^{2N}} d^2\xi_1 \dots d^2\xi_N$$
$$\times \prod_{i<j}^{N} \left(|\xi_i - \xi_j|^{\frac{(\rho_i^\mu, \rho_j^\mu)}{\pi\alpha'}} \right) \tag{4.26}$$

---

[1] If $\Delta(\xi) = \int d^D k \cdot \frac{e^{ik\xi}}{|K|^\beta}$ with $\beta \in R$, then on the dimensional regularization scheme $\Delta(0) = \int d^D k |K|^{-\beta} = 0$, even if $\beta = 0$.

4.3 The Nambu-Goto Extrinsic Path String

Another important classical local diffeomorphism invariant surface functional, closely related to the Q.C.D. ($SU(\infty)$) string ([1])l, is the so-called extrinsic functional which is defined by the square of the surface mean curvature (see eq. (4.9)), with $\gamma \in R$

$$\mathcal{F}_{\text{exta}}(S) = \gamma^2 \int_\Omega [\sqrt{h}\,(\Delta_h X_\mu)^2](\xi) d^2\xi. \tag{4.27}$$

Let us add suh functional to the area functional eq. (4.1) and consider the associated string path integral propagator eq. (4.4) (in the trivial topological secta of surface S_c).

$$\begin{aligned} \overline{G}(C) = \int_{X_\mu(\xi)|_{\xi\in\partial\Omega}=C^\mu(\sigma)} & d_h\mu[X_\mu(\xi)] \\ & \times \exp\left\{-\frac{1}{2\pi\alpha'}\int_\Omega (h(X_\mu(\xi)))^{1/2} d^2\xi\right\} \\ & \times \exp\left\{-\gamma^2 \int_\Omega [\sqrt{h}(\Delta_h X_\mu)^2](\xi) d^2\xi\right\}. \end{aligned} \tag{4.28}$$

Let us note that even by using the light-cone string coordinate system, One still has a non-polinomial interacting quantum two-dimensional $SO(D)$ scalar field theory.

Namelly

$$\begin{aligned} \overline{\overline{G}}(C) = \int & d_h\mu[X'_\mu(z,\bar{z})] \\ & \times \exp\left\{-\frac{1}{\pi\alpha'}\int_\Omega dz d\bar{z}\left(\frac{1}{2}(\partial_+ X'^\mu)(\partial_- X'^\mu)\right)(z,\bar{z})\right\} \\ & \times \exp\left\{-\gamma^2 \int_\Omega dz d\bar{z}\left[\frac{(\partial_+^2 X'^\mu)(\partial_-^2 X'_\mu)}{(\partial_+ X'^\mu)(\partial_- X'^\mu)}\right]\right\} \\ & \times \delta^{(F)}[(\partial_+ X'^\mu \cdot \partial_+ X'^\mu)] \\ & \times \delta^{(F)}[(\partial_- X'^\mu \cdot \partial_- X'^\mu)]. \end{aligned} \tag{4.29}$$

However one can evaluate formally scattering amplitudes as done previously (see eq. (4.23)) by considering the string quantum trajectory as a small perturbation around the flat metric. The outcome is the following one-loop string scalar scattering amplitude compare with the Nambu-Goto case of the previous section eq. (4.23)

$$A(\rho_1^\mu,\dots,\rho_N^\mu)=\frac{1}{Z}\Bigg\{\int D^F[X_\mu(\xi)] \times\exp\left[-\frac{1}{\pi\alpha'}\int_{R^2}\frac{1}{2}(\partial X_\mu\partial X_\mu)(\xi)d^2\xi\right] \times\exp\left[-\gamma^2\int_{R^2}(\partial^2X_\mu)(\partial^2X_\mu)(\xi)d^2\xi\right] \times\left[\int_{R^{2N}}d^2\xi_j\left(\prod_{j=1}^N\exp(i\rho_\mu^jX_\mu)(\xi_j)\right)\right]\Bigg\}. \tag{4.30}$$

Since eq. (4.30) is a Gaussian Functional Integral, One obtains an exactly result, where we have used the result

$$\left[\frac{1}{(\pi\alpha')}(-\partial^2)+\gamma(\partial^2)^2\right]^{-1}(\xi,\xi1) := \pi\alpha'\left[\left(-\frac{1}{2\pi}\ln|\xi-\xi'|\right)-\frac{1}{2\pi}\left(K_0\left(\left(\frac{1}{\pi\alpha'\gamma^2}\right)^{1/2}|z-z'|\right)\right)\right]. \tag{4.31}$$

Namelly:

$$A(\rho_1^\mu,\dots,\rho_N^\mu) =\int_{R^{2N}}d^2\xi_1\dots d^2\xi N \times\left(\prod_{i<j}^N(\xi_i-\xi_j)^{\frac{(\rho_i^\mu,\rho_j^\mu)}{\pi\alpha'}}\right) \times\left(\exp\left[-\frac{\alpha'}{2}\sum_{i<j}^N K_0\left(\left(\frac{1}{\pi\alpha'\gamma^2}\right)^{1/2}|\xi_i-\xi_j|\right)\right]\right). \tag{4.32}$$

Unfortunatelly, it appears that the corrections coming from the string extrinsic functions do not modify the usual Veneziano-Virassoro bosonic closed string spectrum. But a clear proof of this no go result is still missing.

At this point we analyze the vacuum string energy for the domain $\Omega_{R,T)}$ being a rectangle of sizes $0\le\xi_1\le R$; $0\le\xi_2\le T$ through a one-loop approximation inthe Regge slope constant α' (see eq. (4.15-a) of the previous section).

In this case we have the exact result

$$\overline{G}[C_{(R,T)}] = \exp\left\{-\frac{1}{2\pi\alpha'}RT\right\} \times \left[\det_{\Omega(R,T)}\left(\gamma^2(-\partial^2)\left(\frac{1}{\gamma^2}+(-\partial^2)\right)\right)\right]^{-\frac{(D-2)}{2}}. \quad (4.33)$$

The evaluation of the fourth-order elliptic operator (with Dirichlet cnditions) can be accomplished through the result

$$\left[\det_{\Omega(R,T)}\left(\gamma^2(-\partial^2)\left(\frac{1}{\gamma^2}+(-\partial^2)\right)\right)\right]^{-\frac{(D-2)}{2}} = \left[\left(\det_{\Omega(R,T)}^{-\frac{(D-2)}{2}}[(-\partial^2)]\right) \times \left(\det_{\Omega(R,T)}^{-\frac{(D-2)}{2}}\left[\frac{1}{\gamma^2}+(-\partial^2)\right]\right)\right]. \quad (4.34)$$

But functional determinants have been evaluated in the literature ([7]). (See eq. ([7])). The second order massive operator has the following result

$$\det^{-\frac{(D-2)}{2}}\left[(-\partial^2)+\frac{1}{\gamma^2}\right] = \exp\left\{\frac{(D-2)\pi T}{R}\left(\frac{1}{6}-\frac{R}{\gamma^2 2\pi} + \left(\frac{R}{2\pi\gamma^2}\right)^2 \ln\left(\frac{4\pi e^{-\hat{\gamma}}}{R}\right) + \frac{1}{2}\left(\frac{R}{2\pi\gamma^2}\right)^4\left(\int_0^1 dx(1-x)\sum_{n=1}^{\infty}\left(n^2+x\left(\frac{R}{2\pi\gamma^2}\right)^2\right)^{-\frac{3}{2}}\right)\right\} \times \left\{\prod_{n=-\infty}^{+\infty}\left[\left[1-\exp\left(\left(-\frac{2\pi T}{R}\sqrt{n^2+\left(\frac{R}{2\pi\gamma^2}\right)^2}\right)\right)\right]\right]\right\}^{-(D-2)}. \quad (4.35)$$

Let us remark that $\left(t = \frac{R}{2\pi\gamma^2}\right)$ (appendix B)

$$\int_0^1 dx(1-x)\left[\frac{1}{\left(n^2 + x\left(\frac{R}{2\pi\gamma^2}\right)^2\right)^{3/2}}\right]$$

$$= -\frac{2}{t^2}\left(\frac{1}{(n^2+t^2)^{1/2}}\right) + \frac{2n^2}{t^4(n^2+t^2)^{1/2}}$$

$$+\frac{2}{t^2 n} - \frac{2}{t^4}(n^2+t^2)^{1/2}. \tag{4.36}$$

which unfortunatelly leads to an – in principle – divergence on the string path integral $\overline{G}(C_{R,T})$ for the extrinsic string. However it is possible to get a somewhat phenomenological finite result for the vacuum energy (the interquark potential) for large R $(R \to \infty)$

$$E_{\text{Vac}}(R) = \lim_{T\to\infty}\left\{-\frac{1}{T}\ln G(C_{(R,T)})\right\}. \tag{4.37}$$

The point is that for large R $(R \to \infty)$, the Epstein sum below written has a finite asymptotic behavior (see appendix A for a detailed evaluation), $s \in R$

$$S_{\text{epst}}(s,a^2) := \sum_{n=1}^{\infty}\left(\frac{1}{(n^2+a^2)^s}\right)$$

$$= \frac{1}{\Gamma(s)}\left(\int_0^\infty du U^{s-1}\cdot e^{-U(\lambda t^2)}\left[Tr_{\left\{\begin{matrix} C^2([0,1]) \\ f(0)=f(1)=0\end{matrix}\right\}}\left(e^{U(-\frac{d^2}{dz^2})}\right)\right]\right)$$

$$\overset{a\to\infty}{\sim}\left(\frac{1}{4\pi\Gamma(s)}\right)\left[(a^2)^{-(S-\frac{3}{2}+1)}\right]. \tag{4.38}$$

As a consequence of the finite behavior eq. (4.38), One has the asymptotic (distributional L. Schwartz sense)

$$\sum_{n=1}^{\infty} \frac{1}{\left(n^2 + x\left(\frac{R}{2\pi\gamma^2}\right)^2\right)} \overset{R\to\infty}{\sim} \frac{1}{2(\pi)^{3/2} x \left(\frac{R}{2\pi\gamma^2}\right)^2} . \tag{4.39}$$

After substituting this (formal) result on eq. (4.35), by taking into account the (regularized) integration:

$$\lim_{R\to\infty}\left\{\frac{1}{2}\left(\frac{R}{2\pi\gamma^2}\right)^4\left[\int_0^1 dx(1-x)\left(\sum_{n=1}^{\infty}\frac{1}{(n^2+x\left(\frac{R}{2\pi\gamma^2}\right)^2)^{3/2}}\right)\right]\right\}$$

$$= \frac{1}{2}\left(\frac{R}{2\pi\gamma^2}\right)^4\left[\int_0^1 dx(1-x)\frac{1}{2\pi^{3/2}x\left(\frac{R}{2\pi\gamma^2}\right)^2}\right]$$

$$\sim \frac{\left(\frac{R}{2\pi\gamma^2}\right)^2}{4\pi^{3/2}}\left(\int_{\varepsilon_{QCD}}^1 \frac{dx(1-x)}{x}\right)$$

$$= \frac{\left(\frac{R}{2\pi\gamma^2}\right)^2}{4\pi^{3/2}}\left[-\ell y(\varepsilon_{QCD})-1\right]$$

$$- -\left\{\left(\frac{(\ell y(\varepsilon_{QCD})+1)}{16\pi^{5/2}\gamma^2}\right)\right\} R^2. \tag{4.40}$$

Here ε_{QCD} is the underlying cut-off on the "Feynman" parameters x ($\varepsilon_{QCD} \le x \le 1$). It is worth remark that the regularization parameters will be absorbed in the bare extrinsic coupling constant γ^2 (somewhat related to the $Q.C.D(SU(\infty))$ coupling constant $\gamma^2 = (g_\infty)^2(\langle 0|F^2|0\rangle_{SU(\infty)})$ (see chapter 1). Namelly:

$$\frac{1}{\gamma^2_{\text{bare}}}(\varepsilon) = \frac{1}{\gamma^2_{\text{bare}}}((\ln \varepsilon_{QCD})+1).$$

The contribution of this term to the phenomenological "interquark potential" on the extrinsic string theory is exactly given below (to be added to the pure Nambu-Goto term eq. (4.18).

$$V^{\text{extrinsic}}(R) = -\left(\frac{(D-2)}{6}\cdot\frac{\pi}{R}\right) + \left(\frac{(D-2)}{2}\cdot\frac{1}{\gamma_{\text{ren}}^2}\right)$$

$$-\left(\left(\frac{(D-2)}{4\pi}\cdot\frac{1}{\gamma_{\text{ren}}^4}\right) R\cdot\ln(4\pi e^{-\hat{\gamma}})\right)$$

$$+\left(\frac{(D-2)}{4\pi}\left(\frac{1}{\gamma_{\text{ren}}^4}\right)(R\ln R)\right)$$

$$+\left(\frac{(D-2)}{16\pi^{5/2}}\frac{1}{\gamma_{\text{ren}}^2}\cdot R\right). \tag{4.41}$$

We think that eq. (4.41) is an important result on applications of string representations for QCD, since it leads to a growing force that goes to infinite for $R \to \infty$, confining the static color quarks charge, certainly a real "quark confinement", opposite to the pure Nambu-Goto case with a "weak confinement" by a *constant* force. quarks are really confined if a string for QCD holds true at the limit of large R. So a QCD point particle description should only be phenomenological.

We comment also that such hind of behavior for the interquark potential is compatible with a Mandelstan Gluonic propagator of the form ([8])

$$D_m(x-y) = \frac{1}{(2\pi)^D}\left(\int d^D p\cdot e^{ip(x-y)}[(\ell y(|p|^2))/|p|^4]\right). \tag{4.42}$$

4.4 Studies on the perturbative evaluation of closed Scattering Amplitude in a Higher order Polyakov's Bosonic String Model

It is well known that Polyakov covariant string theory with exactly soluble Liouville two dimensional model (the famous non critical string) has in its protocol the main hope of suppressing the tachionic excitation of the usual Nambu-Goto string ([10], [11], [12])[2].

By the other side it is less known that Polyakov-Liouville effective string theory is somewhat phenomenological in the sense that it has already built in the assumption of "weak" two dimensional induced quantum gravity, since one replaces its covariant path integral measure $D^F[e^{\varphi/2}] \equiv$

[2]For a detailed presentation of the Polyakov's string path integral see §3 of [9].

$\prod_{\xi\in D} d(e^{\varphi/2}(\xi))$ by the usual Feynman path measure $D^F[e^{\varphi/2(\xi)}] \sim D^F[\varphi(\xi)]$ when considering the final effective Liouville-Polyakov field theory on string world-sheet.

So in this approximate scheme of non critical strings (but quite useful ([2])), we in this section (somewhat pedagogical) introduce a somewhat toy model of higher-order two dimensional Polyakov covariant string with improved ultra violet behavior and show the exactly solubility of the resulting covariant Polyakov path integrations.

Let us start our analysis by considering the theory's N-point off-shell closed Scattering amplitude defined by the following Polyakov's bosonic string general path integral

$$A(P_1^\mu,\ldots,P_N^\mu) = \int d^{\text{cov}}\mu[g_{ab}(\xi)]e^{-\frac{\mu^2}{2}\int_{R^2}(\sqrt{g})(\xi)d^2\xi}$$

$$\times\left[e^{\overbrace{-\frac{\gamma}{2}\int_{R^2} d^2\xi d^2\xi'\sqrt{g(\xi)}\sqrt{g(\xi')}R(\xi)K_g(\xi,\xi')R(\xi')}^{=\text{newproposed term}}}\right]$$

$$\times\left[\int d^{\text{cov}}\mu[X^\mu(\xi)]\exp\left\{-\frac{1}{2}\int_{R^2}(\sqrt{g}\,g^{ab}\partial_a X^\mu\partial_b X^\mu)(\xi)d^2\xi\right\}\right]$$

$$\times\left\{\int_{R^{2N}}\prod_{i=1}^N\left[\sqrt{g(\xi_i)}\cdot\exp(P_i^\mu X_\mu(\xi_i))\right]d^2\xi_i\right\}. \tag{4.43}$$

Here the covariant functional measures are the functional volume elements associated to the functional Riemann metrics ([1], [2]. [3])

$$||\delta g_{ab}||^2 = \int_{R^2}\left[\sqrt{g}((\delta g_{ab})(g^{aa'}g^{bb'} + cg^{ab}g^{a'b'})(\delta g_{a'b'}))\right]d^2\xi$$

$$||\delta X^\mu||^2 = \int_{R^2}[\sqrt{g}\,\delta X^\mu\delta X^\mu](\xi)d^2\xi. \tag{4.44}$$

The Green function $K_g(\xi,\xi')$ of the Laplace-Beltrami operator on the presence of the metric $g_{ab}(\xi) = e^{\varphi(\xi)}\delta_{ab}$, in the conformal gauge, with a covariant cut-off already built in is given explicitly by the Riesz-Hadamand formula ([4], [5]) on R^2

$$K_\varphi(\xi,\xi') = \begin{cases} -\dfrac{1}{2\pi}\ell y|\xi-\xi'| & \xi \neq \xi' \\ \\ \dfrac{1}{2\pi}\left(\dfrac{1}{\varepsilon}\right) - \dfrac{1}{2\pi}\varphi(\xi) - \dfrac{1}{2\pi}\ell y(\varepsilon) & \xi = \xi' \end{cases} \tag{4.45}$$

Note that eq. (4.3) is a direct result that in the conformal gauge $g_{ab} = e^{\varphi(\xi)}\delta_{ab}$ the Laplace-Beltrami operator reduces to the usual Laplace operator for $\xi \neq \xi'$. And for those points $\xi = \xi'$, one should use the famous parameter formula of J. Hadamard for the Laplace Beltrami Operator on $\mathbb{R}^2$ togheter by taking into account the fact that the geodesic distance $S(\xi,\xi')$ in R^2, endowed with a metric in the conformal gauge is exactly given by $S(\xi,\xi') = e^{\varphi(\xi)}|\xi-\xi'|$ for $\xi \to \xi'$.

In other words, for $\xi \to \xi'$, we have the regularized asymptotic behavior an R^N ([5])

$$\begin{aligned} K(\xi,\xi') &\sim \lim_{N\to 2}\left\{\frac{\Gamma(\frac{1}{2}N)}{2\pi^{(\frac{N}{2})}(N-2)} e^{(2-N)\ell g(S(\xi,\xi'))}\right\} \\ &\overset{\substack{N\to 2\\ \varepsilon\to 0}}{\sim} \frac{1}{2\pi}\left(\frac{1}{N-2}\right) - \frac{1}{2\pi}\ell g\left(e^{\varphi(\xi)}(|\xi-\xi'|^2+\varepsilon^2)^{1/2}\right) \\ &\overset{|\xi-\xi'|\to 0}{\sim} \frac{1}{2\pi\varepsilon} - \frac{\varphi(\xi)}{2\pi} - \frac{1}{2\pi}\ell y(\varepsilon). \end{aligned} \tag{4.46}$$

In order to evaluate perturbativelly eq. (4.1) around the flat metric $g_{ab}(\xi) \sim \delta_{ab}$, we consider the following approximation on the resulting effective functional measure on the metric field by replacing it by its weak metric feynman measure

$$
\begin{aligned}
d\mu[g_{ab}] &= D^F[e^{\frac{\varphi}{2}(\xi)}] \\
&\times \exp\left\{-\frac{26}{48}\int_{R^2_+}\frac{1}{2}(\partial\varphi)^2(\xi)d^2\xi\right\} \\
&\times \exp\left\{-\frac{\mu^2}{2}\int_{R^2_+} e^{\varphi(\xi)}d^2\xi\right\} \\
&\times \exp\left\{-\frac{\gamma^2}{2}\int_{R^2}(\partial^2\varphi)^2(\xi)d^2\xi\right\} \\
&\overset{\text{(replace by)}}{\sim} D^F[\varphi(\xi)]\exp\left\{-\frac{26}{48}\int_{R^2}\frac{1}{2}(\partial\varphi)^2 d^2\xi\right\} \\
&\times \exp\left\{-\frac{\gamma^2}{2}\int_{R^2}(\partial^2\varphi)^2(\xi)d^2\xi\right\} \\
&\times \exp\left\{-\frac{\mu^2}{2}\int_{R^2} e^{\varphi(\xi)}d^2\xi\right\}. \qquad (4.47)
\end{aligned}
$$

Note that we have further replaced the fourth-order (non-local) $2D$ gravity term by the simple local result exposed below

$$
\begin{aligned}
&\exp\Big\{-\frac{\gamma}{2}\int d^2\xi d^2\xi' e^{\varphi(\xi)}e^{\varphi(\xi')}\Big(\overbrace{e^{-\varphi(\xi)}\partial^2\varphi(\xi)}^{R(\xi)}\Big) \\
&\times \frac{\delta^{(2)}(\xi-\xi')}{e^{\varphi(\xi)}}(^{-\varphi(\xi')}\partial^2\varphi(\xi))\Big\} \\
&= \exp\left\{-\frac{\gamma}{2}\int d^2\xi d^2\xi'(\partial^2\varphi(\xi))\frac{\delta^{(2)}(\xi\quad\zeta')}{e^{\varphi(\xi)}}(\partial^2\varphi)(\xi')\right\} \\
&\sim \exp\left\{-\frac{\gamma}{2}\int d^2\xi(\partial^2\varphi)^2(\xi)\right\}. \qquad (4.48)
\end{aligned}
$$

Here we have substituted the covariant Laplace Beltrami operator Green Function by its mathematical distributional limit $\delta^{(2)}_{\text{cov}}(\xi-\xi')$, but for a back ground "weak metric" $g_{ab}(\xi)=\delta_{ab}$.

In other words:

$$
\frac{\delta^{(2)}(\xi-\xi')}{e^{\varphi(\xi')}} \sim \frac{\delta^{(2)}(\xi-\xi')}{1+\varphi(\xi')+\dots} \sim \delta^{(2)}(\xi-\xi').
$$

It is worth to call the reader attention that we have preserved in its integrity the Liouville cosmological term $\exp\left\{-\frac{\mu^2}{2}\int_{R^2} e^{\varphi(\xi)}d^2\xi\right\}$, since the main interest in this section is to present perturbative calculations on the

N-point off-shell closed scattering amplitude that show us that such important term does not affects its pole singularities of those usual closed string scattering amplitudes Kobo-Milsen-Virasoro result for $\mu^2 = 0$; even in presence of metric fourth-order term as proposed by ours.

After collecting all the above results and remarks, we are lead to evaluate perturbativelly on the Liouville-Polyakov cosmological constant μ^2, the following standard Liouville-Polyakov path integral

$$
\begin{aligned}
\tilde{A}(P_1, \dots, P_N) = \int D^F[\varphi(\xi)] \exp\left\{-\frac{(26-d)}{48\pi} \int_{R^2} \frac{1}{2}(\partial\varphi)^2(\xi) d^2\xi\right\} \\
\times \exp\left\{-\frac{\mu^2}{2} \int_{R^2} e^{\varphi(\xi)} d^2\xi\right\} \\
\times \exp\left\{-\frac{\gamma^2}{2} \int_{R^2} (\partial^2\varphi)^2(\xi) d^2\xi\right\} \\
\times \left\{\left[\prod_{i=1}^{N} e^{\varphi(\xi_i)}\right] \left[\prod_{i=1}^{N} e^{i(P_i^\mu \cdot X_\mu(\xi_i))}\right]\right\}.
\end{aligned} \tag{4.49}
$$

By taking into account the momentum conservation on the scattering of the string excitations $\left(\sum_{i=1}^{N}(P_\mu^i)^2 \equiv 0\right)$, One obtains the following Liouville path integral without the ε-covariant cut-off

$$
\begin{aligned}
\tilde{A}(P_1, \dots, P_N) = \int_{R^{2N}} d^2\xi_1 \dots d^2\xi_N \Big(\exp - \Big[\sum_{\substack{i,j=1 \\ (i \neq j)}} (P_\mu^i \cdot P_\mu^j) \\
\times \left(-\frac{1}{4\pi} \ln|\xi_i - \xi_j|^2\right)\Big]\Big) \\
\times \left\{\int D^F[\varphi(\xi)] \exp\left[-\frac{(26-D)}{48\pi} \int_{R^2} \frac{1}{2}(\partial\varphi)^2(\xi) d^2\xi\right]\right. \\
\times \exp\left[-\frac{\gamma^2}{2} \int_{R^2} (\partial^2\varphi)^2(\xi) d^2\xi\right] \exp\left[-\frac{\mu^2}{2} \int_{R^2} e^{\varphi(\xi)} d^2\xi\right] \\
\left. \times \left(\exp\left[\left(\sum_{i=1}^{N} \varphi(\xi_1)\right) - \left(\sum_{i=1}^{N} (P_\mu^i)^2 \frac{\varphi(\xi_i)}{4\pi}\right)\right]\right)\right\}.
\end{aligned} \tag{4.50}
$$

It is worth to point out that the zeroth order $2D$ quantum metric $g_{ab}(\xi) = \delta_{ab}$ $(\varphi(\xi) = 0)$, the Liouville path integral eq. (4.8) produces a scattering amplitude structurally similar to the famous Veneziane-Koba-Nielsen-Virasoro closed scattering amplitude (in the Euclidean space-time

R^N)

$$\tilde{A}_{(0)}(P_1,\dots,P_N)$$

$$= \int_{R^{2N}} d^2\xi_1 \dots d^2\xi_N \left(\exp - \left[\sum_{\substack{i,j=1 \\ i\neq j}} (P^i_\mu \cdot P^j_\mu) - \frac{1}{4\pi} \ln |\xi_i - \xi_j|^2 \right] \right)$$

$$= \int_{R^{2N}} d^2\xi_1 \dots d^2\xi_N \left(\prod_{\substack{(i,j=1) \\ i\neq j}} \left[|\xi_i - \xi_j|^{\frac{(P^i_\mu \cdot P^j_\mu)}{2\pi}} \right] \right). \tag{4.51}$$

The key point of our study is the exactly path-integral evaluation of the fourth-order improved Liouville path integral in a perturbative expansion in the Polyakov's cosmological constant μ^2. Namelly

$$\begin{aligned} F_{\text{Liouville}}(P^i_\mu) \equiv & \int D^F[\varphi(\xi)] \exp\left[-\frac{(26-D)}{48\pi} \int_{R^2} \frac{1}{2} (\partial\varphi)^2(\xi) d^2\xi \right] \\ & \times \exp\left[-\frac{\gamma^2}{2} \int_{R^2} (\partial^2\varphi)^2(\xi) d^2\xi \right] \\ & \times \left\{ \sum_{n=0}^{\infty} \left(-\frac{\mu^2}{2} \right)^n \int_{R^{2N}} e^{\varphi(z_1)} \dots e^{\varphi(z_N)} d^2z_1 \dots d^2z_N \right\} \\ & \times \left(\exp\left[\sum_{i=1}^{N} \varphi(\xi_i)(1 - \frac{(P^i_\mu)}{4\pi}) \right] \right). \end{aligned} \tag{4.52}$$

Since the above written path integral is a fourth-order Gaussian functional integral, with correlation function given explicitly by

$$\left[\frac{26-D}{48\pi} (-\partial^2) + \gamma^2 (\partial^2)^2 \right]^{-1} (\xi, \xi') \equiv \mathcal{L}^{-1}(\xi.\xi')$$

$$\equiv \frac{48\pi}{26-D} \left[\left(-\frac{1}{2\pi} \ln |\xi - \xi'| \right) - \frac{1}{2\pi} \left(K_0 \left(\sqrt{\frac{26-d}{48\pi\gamma^2}} |\xi - \xi'| \right) \right) \right], \tag{4.53}$$

one obtains the result below, with the path integral average Notation,

$$\langle \mathcal{O}(\varphi) \rangle_\varphi = \frac{1}{2} \left\{ \int D^F[\varphi] e^{-\frac{1}{2} \int \varphi(\xi) \mathcal{L}_\xi \varphi(\xi) d^R\xi} \mathcal{O}(\varphi) \right\}. \tag{4.54}$$

Namely: (for each perturbative order M in μ^2)

$$F^{(M,N)}_{\text{Liouville}}(P^i_\mu) = \left\langle \prod_{j=1}^{N}\prod_{i=1}^{M} \int_{R^{2N}} < e^{\varphi(z_1)}\dots e^{\varphi(z_n)} \times e^{\varphi(\xi_i)\left(1-\frac{(P^i_\mu)^2}{4\pi}\right)} \right\rangle_{\varphi}$$

$$= \left\{ \left(\prod_{(t,s=1)}^{N} \exp\left[+\frac{1}{2}\left(1-\frac{(P^t_\mu)^2}{4\pi}\right)\left(1-\frac{(P^s)^2}{4\pi}\right)\mathcal{L}^{-1}(\xi_t,\xi_s)\right]\right)\right\}$$

$$\times \left(\prod_{(h,q)}^{M} \exp\left[+\frac{1}{2}\mathcal{L}^{-1}(z_h,z_q)\right]\right)$$

$$\times \left(\prod_{t=1}^{N}\prod_{q=1}^{M} \exp\left[+\frac{1}{2}\left(1-\frac{(P^q_\mu)^2}{4\pi}\right)\mathcal{L}^{-1}(z_q,z_t)\right]\right). \quad (4.55)$$

We remark now that the singularities type pole (paticle string excitation that would modify the zeroth-order Veneziano-Kobe-Nielsen-Virasoro closed bosonic string dual model will come from the "pinch" ultra-violet singularities $\xi_t \to \xi_s$! However from the ultra-violet behavior below depicted

$$\lim_{\xi_t\to\xi'_s} \frac{1}{2\pi} K_0\left(\sqrt{\frac{26-D}{48\pi\gamma^2}}\,|\xi_t-\xi'_s|\right)$$

$$= \lim_{\xi_t\to\xi'_s}\left[+\frac{1}{2\pi}\ln|\xi_t-\xi'_s| + \frac{1}{2\pi}\ln\left(\sqrt{\frac{26-D}{96\pi\gamma^2}}\right) - \frac{\psi(1)}{2\pi}\right], \quad (4.56)$$

one can see that there is the cancealling of the logaritmic terms on eq. (4.12) for the "pinch" points protocol to analyze poles on string scattering amplitudes.

As a result, at each perturbative order M, these terms are expected to not contribute to the bosonic closed string spectrum eq. (4.10).

That is the main conclusion of our section: the Polyakov proposal of a quantum bosonic string as a theory of D (ill defined!) two-dimensional massless scalar fields interacting with two-dimensional induced Liouville-Polyakov quantum gravity, (expected to be computationally effective for "weak" two-dimensional quantum gravity), still does not alter the usual (tachionic) old closed bosonic string of the Veneziano-Virasoro dual model, even with a improved ultra-violet behavior as proposed by ours.

It appears thus that the introduction of induced quantum gravity in the Polyakov proposal, could not alter the usual tackionic spectrum of the bosonic string, even if one considers a somewhat "higher order" Polyakov-Liouville induced $2D$ quantum gravity. As a result, one is naturally lead to consider further fermionic (intrinsic, extrinsic or supersymetric) degrees of freedom on the string world sheet ([1], [6]), in oder to have candidate to be physically sensible (see chapter 3).

4.5 Appendix A: The distributional limit of the Epstein function

Let us try to evaluate (in some yet undiscovered asymptotic distributional theory) the limit of large a of the so called Epstein function for $S \in R$

$$
\begin{aligned}
S_{\text{epstein}}(s,a^2) &= \sum_{\mu=1}^{\infty} \frac{1}{(\mu^2+a^2)^s} \\
&= \frac{1}{\Gamma(s)} \left(\int_0^\infty U^{s-1} e^{-a^2U^2} e^{-Un^2} dU \right) \\
&= \frac{1}{\Gamma(s)} \left(\int_0^\infty U^{s-1} \cdot e^{-a^2U^2} \cdot Tr_{\left\{ \begin{matrix} C^2([0,1]) \\ \text{Dirichlet} \end{matrix} \right\}} \left(e^{-U\left(\frac{-d^2}{d\xi'}\right)} \right) \right).
\end{aligned} \tag{A-1}
$$

For $a \to \infty$,k certainly $U \to 0$ on the U-integrand, and the Seeley asymptotic expansion for the second-order operator $-\frac{d^2}{dz^2}$ holds true.

Theta is:

$$
\begin{aligned}
S_{eps}(s,a^2) &\overset{a^2\to\infty}{\sim} \frac{1}{\Gamma(s)} \left[\int_0^\infty dU \cdot U^{s-1} e^{-Ua^2} \frac{1}{4\pi U^{1/2}} \right] \\
&\overset{a^2\to\infty}{\sim} \frac{1}{4\pi\Gamma(s)} \left[\int_0^\infty dU \cdot U^{s-\frac{3}{2}} e^{-Ua^2} \right] \\
&\overset{a^2\to\infty}{\sim} \frac{1}{4\pi\Gamma(s)} \left[\frac{\Gamma(s-\frac{3}{2}+1)}{(a^2)^{s-\frac{3}{2}+1}} \right]
\end{aligned}
$$

4.6 Appendix B: Integral Evaluation

Let us elementarily evaluate the following integral

$$\int_0^1 dx(1-x)\frac{1}{(a^2+xb)^{3/2}} = \int_0^1 \frac{dx}{(a^2+xb)^{3/2}} - \int_0^1 dx\frac{x}{(a^2+xb)^{3/2}}$$

$$\overset{(x=\frac{v-a^2}{b})}{=} \frac{1}{b}\left(\int_{a^2}^{a^2+b} dv\cdot v^{-\frac{3}{2}}\right) - \frac{1}{b^2}\left(\int_{a^2}^{a^2+b} dv\cdot v^{-\frac{1}{2}}\right)$$

$$-\frac{a^2}{b^2}\left(\int_{a^2}^{a^2+b} dv\cdot v^{-\frac{3}{2}}\right)$$

$$= \left[(a^2+b^2)^{-\frac{1}{2}}\left(-\frac{2}{b}+\frac{2a^2}{b^2}\right)\right]$$

$$+\left[\frac{2}{ab}+\frac{2a}{b^2}-\frac{2a}{b^2}\right] - \frac{2}{b^2}(a^2+b)^{1/2}$$

$$= \left(-\frac{2b+2a^2}{b^2}\right)\frac{1}{(a^2+b)^{1/2}} + \frac{2}{ab} - \frac{2}{b^2}(a^2_+b)^{1/2}. \tag{B-1}$$

As a consequence ($\mu^2 = n^2$)

$$\int_0^1 dx(1-x)\frac{1}{(\mu^2+xa^2)^{3/2}} = -\left(\frac{2}{a^2}\right)\left(\frac{1}{(\mu^2+a^2)^{1/2}}\right)$$

$$+\frac{1}{a^4}\cdot\frac{2\mu^2}{(\mu^2+a^2)^{1/2}}$$

$$+\frac{2}{a^2\mu} - \frac{2}{a^4}(\mu^2+a^2)^{1/2}. \tag{B-2}$$

4.7 Appendix C: On the perturbative evaluation of the bosonic string closed scattering amplitude on Polyakov's framework

The fundamental observable on Polyakov's bosonic string theory ([1], [2]) is the closed scattering amplitude which is given explicitly by the covariants

$2D$ induced quantum gravity path integral in moments space

$$\tilde{A}(P_\mu^1, \dots, P_\mu^N) = \prod_{j=1}^{N} \left[\int d^2\xi_j \times \left(\frac{1}{Z} \int D^F[\bar{\rho}(\xi)] \right.\right.$$

$$\times \exp\left\{ -\frac{(26-D)}{12\pi} \int_{\mathbb{R}^2} \left(\frac{1}{2} \left(\frac{\partial_a \bar{\rho}}{\bar{\rho}} \right)^2 \right)(\xi) d^2\xi - \frac{1}{2}\mu_R^2 \int_{\mathbb{R}^2} \bar{\rho}^2(\xi) d^2\xi \right\}$$

$$\times \exp\left\{ \sum_{j=1}^{N} 2\ln(\bar{\rho}(\xi_j)) \right\}$$

$$\left.\left. \times \exp\left\{ -\sum_{(i,j)=1}^{N} (P_\mu^i \cdot P_\mu^j)_{R^N} K^{(\varepsilon)}(\xi_i, \xi_j) \right\} \right) \right]. \tag{C-1}$$

Here the covariant regularized Polyakov's Green function is given explicitly by eq. (4.3) (with the identification $(\bar{\rho}(\xi))^2 = e^{\varphi(\xi)}$).

After substituting eq. (4.3) into eq. (C-1), one obtain the outcome expressed now as a sigma model like perturbativelly renormalizable path integral defined by a trully Feynman measure $D^F[\bar{\rho}(\xi)]$ ([1], [9]).

We have thus to evaluate in term of the $\frac{1}{D}$-expansion, the natural perturbative parameter expansion on two-dimensional quantum gravity applied for quantum strings with domain parameter being $\mathbb{R}^2$ (closed string) (see footnote)

$$\bar{\rho}(\xi) = 1 + \left(\frac{12\pi}{26-D} \right)^{1/2} h(\xi). \tag{C-2}$$

Here we have considered our classical background metric, the flat metric $g_{ab}(\xi) = 1\delta_{ab}$ on $\mathbb{R}^2$, as it should be in Einstein like gravitation theories.

At one-loop order, one arrives at the following path integral (with the normalization factor $Z = 1$).

The full $\frac{1}{D}$ expanded path integral is given by

$$Z = \int D^F[h] \exp\left\{ -\frac{1}{2} \int_{\mathbb{R}^2} \left[(\partial h)^2 \left(\sum_{n=1}^{\infty} (-1)^{n-1} n \varepsilon^{n-1} h^{n-1} \right) \right] (\xi) d^2\xi \right\}$$

$$\times \exp\left\{ -\mu^2 \int_{\mathbb{R}^2} h^2(\xi) d^2\xi \right\} \tag{C-3}$$

with the ortoghonal constraint of the fluctuating piece $h(\xi)$ in relation to the flat $\mathbb{R}^2$ background

$$\int_{\mathbb{R}^2} h(\xi) \cdot 1 \cdot d^2\xi = 0. \tag{C-4}$$

From the usual renormalization power counting interactions of the form $((\partial h)^2 h^m)(\xi)$ have the renormalization index $r = \left(\frac{(D-2)}{2}\right) b + \frac{(D-1)}{2} f + \delta - D$ for an interaction of general for $g(\partial)^\delta \phi^b \psi^F$. In our case $g = \varepsilon$, $D = 2$, $f = 0$, $\delta = 2$, $b = m$, which leads to the theory's renormalizability $r = 0$ in $\mathbb{R}^2$.

$$A(P_\mu^1, \dots, P_\mu^N) = \frac{1}{2}\left\{\int D^F[h] \exp\left(-\frac{1}{2}\int_{R^2}[(\partial h)^2 + \mu^2 h^2](\xi) d^2\xi\right)\right\} \times \left\{(\varepsilon)^{(\sum_{j=1}^N (P_\mu^j)^2)} \times \int_{\mathbb{R}^2} \prod_{j=1}^N d^2\xi_j \left(\prod_{1<j} |\xi_i - \xi_j|^{\frac{(P_\mu^i \cdot P_\mu^j)}{2\pi}\mathbb{R}^N}\right)\right\} \times \left[\prod_{j=1}^N \exp\left(2\left(\frac{12\pi}{26-D}\right)^{1/2}\left(1 - \frac{(P_\mu^j)^2}{4\pi}\right) h(\xi_j)\right)\right]. \tag{C-5}$$

Here we remark the use of the one-loop approximation on the object $\left(\varepsilon = \left(\frac{12\pi}{26-D}\right)^{1/2}\right)$

$$\left[\prod_{j=1}^N (\bar\rho(\xi_j))^{2(1-\frac{(P_\mu^j)^2}{4\pi})}\right] = \left[\prod_{j=1}^N e^{2(1-\frac{(P_\mu^j)^2}{4\pi})\ln(1+\varepsilon h(\xi_j))}\right] \overset{\text{one-loop order}}{\sim} \left\{\prod_{j=1}^N \exp\left[2\varepsilon\left(1 - \frac{(P_\mu^j)^2}{4\pi}\right) h(\xi_j)\right]\right\}. \tag{C-6}$$

At this point one can see that at a perturbativelly renormalized "$\frac{1}{D}$ dx-pansion" ($D \to -\infty$) ([1], [9]), the contribution of the Liouville-Polyakov's dynamics is a multiplication factor as given below ($\sum_{j=1}^N (P_\mu^j)^2 = 0$ for a physical elastic string excitation scattering)

$$\tilde A(P_\mu^1, \dots, P_\mu^N) = \int_{\mathbb{R}^2} d^2\xi_j \left[\prod_{i<j} |\xi_i - \xi_j|^{(\frac{P_\mu^i \cdot P_\mu^j}{2\pi})_{\mathbb{R}^N}}\right] \exp\left\{\sum_{i<j=1}^N \frac{\overbrace{24\pi}^{2\varepsilon^2}}{(26-D)}(1-(P_\mu^i)^2)(1-(P_\mu^j)^2) \times \left(\frac{1}{2\pi} K_0(\mu_R|\xi_i - \xi_j|)\right)\right\}. \tag{C-7}$$

By considering $\varepsilon \to 0$ $(D \to -\infty)$, one re-obtains the well-known A.M. Virasoro-B. Sakita closed scattering from the old dual models.

Another useful remark is that the behavior of eq. (C-5) for $\xi_i \to \xi_j$ is the same of the Virasoro-Sakita result, implying, thus, that by just taking into account the Liouville-Polyakov degree of freedom on Bosonic Srings does not alter the poles of the scattering amplitudes, so leasing to the same drawbacks of the old theory.

We conclude that nonperturbative effects (or others intrinsic degrees of freedom ([1])) must be taken into account to remove the tachion from the Polyakov Bosonic string spectrum ([1]).

4.8 References

[1] Luiz C.L. Botelho, Methods of Bosonic and Fermionic Path Integrals, Nova Science Publisher, Inc., 2009.
[2] A.M. Polyakov, Gauge Fields and Strings, Harwoud Academic Publisher, Chur., (1987).
[3] Michio Kako, Introduction to Superstrings, Springer Verlag, 1988.
[4] Luiz C.L. Botelho, Lecture Notes in Applied Differential Equations of Mathematical Physics, World Scientific, 2008.
[5] M.J. Duff, Partial Differential Equations, Toronto Press, (1956).
[6] M.B. Green, J.H. Schwarz & E. Witten, Superstring theory, vol. 2, Cambridge Monographs on Mathematical Physics, (1996).
[7] C. Itzykson & J.M. Drouffe, Statistical Field theory, vol. 1, Cambridge Monographs on Mathematical Physics, (1991).
[8] Luiz C.L. Botelho, Mod. Physics Letters A, vol. 20, No. 12, (2005).
[9] Luiz C.L. Botelho, ISRN High Energy Physics, vol. 2012, Article ID 674985, doc. 10.5402/2012/674985. (Research Article) – chapter 1.
[10] J. Teschner, Class. Quant. Grav., 18, (2001), R153–R222.
[11] Adel Bilal, F. Ferrari, S. Klevitson, arxi No. 1310.1951v2, [hep-th], (1800–2013).
[12] Luiz C.L. Botelho, Gen Relative Gravity (research article), DOI 10.1007/S, 10714-012-1372-1, (May/11/2012).

Chapter 5

The $D \to -\infty$ saddle-point spectrum analysis of the open bosonic Polyakov string in $R^D \times SO(N)$ – The $QCD(SU(\infty))$ string

In this chapter we investigate the role of the chiral anomaly in determining the spectrum at the saddle-point approximation $D \to -\infty$ of the recently considered Polyakov formulation of bosonic strings moving in $R^D \times G$ with $K = 2$, where G is the group manifold $SO(N)$. The main result is, opposite to the critical dimension, that the spectrum is not sensitive to the model chiral anomaly in the $D \to -\infty$ limit.

5.1 Introduction

There is currently strong interest in the study of strings moving in a compact manifold, mainly connected with the problem of dimensional reduction in superstring theories.

Redlich and Schnitzer have considered a formalism for quantization of closed bosonic strings moving in the space $R^D \times G$ with G being the group manifold associated to the group $U(N)$ [or $SO(N)$]. One important feature of these models is its characterization by a positive integer K which is proportional to the ratio of the radius of the inner compactified dimensions and the Regge slope parameter. The main result obtained by these above-quoted authors was the change in the string's critical dimension as a result of the model chiral anomaly.

The purpose of this chapter is to take a further step in the study of these compactified string models by analyzing the role of the chiral anomaly in determining the string spectrum at the saddle-point limit $D \to -\infty$ (Refs. 10 and 11). As our main result we show that the spectrum for the caqse $K = 2$ does not differ from the case $K = 1$, thus leading to the conclusion that the chiral anomaly does not influence the string spectrum (at least at the natural limit $D \to -\infty$ on Polyakov's framework - chapter 4).

5.2 The non-tachyonic spectrum and scalar amplitudes at $D \to -\infty$

Let us start our analysis by considering the open-bosonic-string action for the group $R^D \times SO(N)$ with $K = 2$ (cf. Ref. 8) generalized to the open-string case: Namely,

$$S[\phi^{(A)}(\xi), e^a_\mu(\xi), \psi^{i,\beta}(\xi), A_\mu(\xi)] = \frac{1}{2}\int_{\mathcal{D}} d^2\xi e(\xi)(g^{\mu\nu}\partial_\mu\phi^{(A)}\partial_\nu\phi^{(A)})(\xi)$$
$$\frac{1}{2}\int_{\mathcal{D}} d^2\xi e(\xi)[\bar{\psi}^{i,\beta}e^\mu_a\gamma_a(\partial_\mu + iA_\mu)\psi^{i,\beta}(\xi)] + \mu_0^2\int_{\mathcal{D}} d^2\xi e(\xi) + \lambda_0\int_{\partial\mathcal{D}} ds. \tag{5.1}$$

The string "surface" is characterized by three fields: first, the usual bosonic vector position field $\phi^{(A)}(\xi)$ $(A = 1,\dots,D)$: second, by a set of $O(N) \times O(2)$ two-dimensional Majorana spinors with components denoted by $\{\psi_1^{i,\beta}(\xi), \psi_2^{i,\beta}(\xi)\}$ $(i = 1,\dots,N;\ \beta = 1,2)$ needed to describe the compactified bosonic coordinates; and finally an $A_\mu - O(2)$ gauge field without kinetic term. The two-dimensional parameter domain $\mathcal{D}$ with boundary $\partial\mathcal{D}$ should be taken in a consistent way with the "bosonic string surface" topology.

The presence of the zweibein $e^\mu_a(\xi)$ $(a = 1,2; \mu = 1,2)$ ensures that the action written above is invariant under general Lorentz and coordinate transformations. The Euclidean (curved) Hermitian $\gamma_\mu(\xi)$ matrices we are using satisfy the relations

$$\begin{aligned}&\{\gamma_\mu, \gamma_\nu\} + (\xi) = 2g_{\mu\nu}(\xi), \gamma^\mu(\xi)\gamma_s = i\frac{\varepsilon^{\mu\nu}}{e(\xi)}\gamma_\nu(\xi),\\ &\gamma_\mu(\xi) = e^a_\mu(\xi)\gamma_a,\ g_{\mu\nu}(\xi) = (e^a_\mu e^b_\nu)(\xi)\delta_{ab},\end{aligned} \tag{5.2}$$

where γ_a are the usual flat-space Dirac matrics.

Dual Green's functions for a scattering of string scalar particles on R^D associated with the $SO(N)$ Euclidean sring theory described by the action Eq. (5.1) can be obtained by taking variational derivatives of the generating functional [$\{\lambda_R\}$ denotes the Hermitian generators of the $SO(N)$ Lie algebra]

$$\Gamma]J_R(x)] = \left\langle \exp\left[-\int_{\mathcal{D}} d^2\xi e(\xi)J_R[\phi^{(A)}(\xi)] \times [\psi_1^i(\xi)(\lambda_R)_{ij}\bar{\psi}_2^j(\xi)]\right]\right\rangle, \tag{5.3}$$

where we have suppressed the $O(2)$ indices and consider the fermionic fields as complex fields. The average $\langle\ \rangle$ is defined as in Ref. 8 or chapter 2, but with the appropriate boundary conditions of Refs. 10–13 and $J_R(x)$ $(R = 1, \ldots, N)$ is an external source with $O(N)$ indices.

Before turning to the computation of the N-point off-shell amplitudes it is instructive to remark that Eq. (5.3) is the string generalization of the analogous formulas in an $SO(N)$ quantum particle dynamics:

$$\begin{aligned} \Gamma[J_R(x)] &= \int_{\varphi_\mu(0)=\varphi_\mu(T)} d[e(\zeta)]d[\varphi_\mu(\zeta)]d[\theta_i(\zeta)]e^{-I[J_R(x)]}, \\ I[J_R(x)] &= m\int_0^T d\xi\left(e(\zeta)\dot{\varphi}_\mu(\zeta)^2 + \frac{1}{e(\zeta)}\right) + i\int_0^T \theta_i(\zeta)\dot{\theta}_i(\zeta)d\zeta \\ &\quad + \int_0^T d\zeta e(\zeta)J_R[\varphi_\mu(\zeta)][\theta_i(\zeta)(\lambda_R)_{ij}\theta_j(\zeta)]. \end{aligned} \tag{5.4}$$

In a gauge $e(\zeta)$ =const, we find that $\lim\limits_{J_R(x)\to 0} \delta^2\Gamma[J_R(x)]/\delta J_{R_1}(x_1)\delta J_{R_2}(x_2)$ is the quantum-mechanical propagator of a free $SO(N)$ bosonic particle.

Let us now pass on to the problem of evaluating the scalar N-point off-shell scattering amplitude through the correlation functions of the associated effective quantum Liouville theory for the case of $\mathcal{D}$ being the upper half-plane $R^2_+ = \{(\xi_1, \xi_2) \mid \xi_2 \geq 0\}$. In momentum space, it is given by

$$\begin{aligned} &\hat{A}_{R_1}, \ldots, R_N(p_1, \ldots, p_N) \\ &= \left\langle \int_{R^2_+} \prod_{j=1}^N d^2\xi_j^{(H)} e(\xi_j) \exp\left[i(p_j^{(A)}; \phi^{(A)}(\xi_j))\right] \left[\psi_1^i(\xi_j)(\lambda_{R_j})_{il}\bar{\psi}_2^l(\xi_j)\right] \right\rangle, \end{aligned} \tag{5.5}$$

where $(\,;\,)$ means the Euclidean scalar product over the Lorentz indices and

$$\prod_{j=1}^N d^2\xi_j^{(H)} = \prod_{j=1;j\neq a\neq b\neq c} d^2\xi_j|\xi_b - \xi_a|^2|\xi_c - \xi_b|^2$$

is the Mbius-invariant Haar measure which takes into account the (physical) residual symmetry of the projective group not fixed by the conformal gauge.

The physical spectrum is determined by considering the poles of the expression $\hat{A}_{R_1} \cdots R_N(p_1, \ldots, p_N)$ and the associated residues identified with the *on-shell* scattering amplitudes for the related string excitation.

Proceeding as in Refs. 10–13 by introducing the family $\mathcal{L}_j = e^{-2(j+1)\delta(Z,Z^*)}\partial_{Z^*}e^{2j\delta(Z,Z^*)}\partial_Z$ of self-adjoint operators acting on an appropriate space of two-component real functions on $R^2_+ = \{Z = \xi_1 + i\xi_2; \xi_2 \geq 0\}$

we can perform the Gaussian functional integration over the scalar fields in the conformal gauge $e^a_\mu(Z,Z^*) = e^{\delta(Z,Z^*)}\delta^a_\mu$. This yields the result

$$\det{}^{-\frac{D}{4}}(\mathcal{L}_0)\exp\left[-\left(\sum_{(i,j)=1}^{N}(p_i^{(A)};p_j^{(A)})K^{(\varepsilon)}(Z_i,Z_j,2\delta(Z_i,Z_i^*))\right)\right], \quad (5.6)$$

where $K^{(\varepsilon)}(Z_i,Z_j,2\delta(Z_i,Z_i^*))$ denotes the Neumann problem Green's function of the covariant Laplacian conformally regularized on the half-plane R^2. Its expression reads.

$$K^{(\varepsilon)}(Z,Z^*,2\delta(Z,Z^*)) = \begin{cases} -\frac{1}{2\pi}\ln(|Z-Z'||Z-Z'^*|), & Z \neq Z', \\ \frac{\delta(Z,Z^*)}{2\pi} - \frac{1}{4\pi}\ln\varepsilon - \frac{1}{2\pi}\ln|Z-Z^*|, & Z = Z'. \end{cases} \quad (5.7)$$

$(e^a_\mu(Z,Z^*) = e^a_\mu(\xi) = \exp[\delta(Z,Z^*)]\delta^a_\mu)$.

Let us now evaluate the functional integrations over the fermionic compactified string degrees of freedom and also of the $U(1)$ gauge field $A_\mu(\xi)$.

Therefore, we face the problem of evaluating the functional integrals

$$\begin{aligned} &W_{R_1}\cdots R_N[e^\mu_a(\xi)](Z_1,\ldots,Z_N) \\ &= \int \mathcal{D}[\psi^i(\xi)]\mathcal{D}[\bar{\psi}^i(\xi)]\mathcal{D}[A_\mu(\xi)]\exp\left(-\frac{1}{2\pi}\int_{R^2_+} d^2\xi[\psi^i\ \not{D}(A_\mu)\bar{\psi}^i](\xi)\right) \\ &\times\left(\prod_{j=1}^{N} psi^i_1(\xi_j)(\lambda_{R_j})_{il}\bar{\psi}^l_2(\xi_j)\right), \end{aligned} \quad (5.8)$$

where $\not{D}(A_\mu) = ie^a_\mu\gamma_a(\partial_\mu + iA_\mu)$ denotes the Dirac operator in the presence of the background $O(2)$ gauge field.

We evaluate Eq. (5.8) by using the fact that due to the chiral anomaly the fermionic functional measures is not invariant under the chiral transformations. Hence, by making the decoupling variable change in Eq. (5.8) (see chapter 3),

$$\psi^i(\xi) = e^{i\gamma_5\eta(\xi)}\chi^i(\xi), \quad \bar{\psi}^i(\xi) = \bar{\chi}^i(\xi)e^{i\gamma_5\eta(\xi)}, \quad A_\mu(\xi) = -\frac{\varepsilon^{\mu\nu}}{e(\xi)}(\partial_\nu\eta)(\xi), \quad (5.9)$$

we obtain the result

$$W_{R_1} \cdots R_N[e^a_\mu(\xi)](Z_1, \ldots, Z_N)$$
$$= (\det \mathcal{L}_0)^{1/2} \int \mathcal{D}[\chi^i(\xi)]\mathcal{D}[\bar{\chi}^i(\xi)]\mathcal{D}[\eta(\xi)]$$
$$\exp\left(-\frac{1}{2}\int_{R^2_+} e(\xi)[\chi^i(i\gamma_a\partial_a)\bar{\chi}^i](\xi)d^2\xi\right)$$
$$\times \exp\left(-\frac{1}{2\pi}\int_{R^2_+} d^2\xi e(\xi)(g^{\mu\nu}\partial_\mu\eta\partial_\nu\eta)(\xi)\right)$$
$$\times \left(\prod_{j=1}^{N}[e^{i\gamma_5\eta(Z_j,Z_j^*)}\chi^i(Z_j,Z_j^*)]_1(\lambda_{R_j})_{il}[\bar{\chi}^l(Z_j,Z_j^*)e^{i\gamma_5\eta(Z_j,Z_j^*)}]_2\right). \tag{5.10}$$

Since in Eq. (5.10) the $(\eta(\xi), \chi^i(\xi), \bar{\chi}^i(\xi))$ fields are decoupled we can evaluate the corresponding functional integrations, yielding (in the conformal gauge)

$$W_{R_1} \cdots R_N[\delta(Z.Z^*)](Z_1, \ldots, Z_N)$$
$$= (\det \mathcal{L}_{-1/2})^{1/4}\left(Tr\sum_{(\sigma)}(\lambda_{R_{\sigma(1)}} \cdots \lambda_{R_{\sigma(N)}})\right)\exp\left(-\sum_{i=1}^{N}\delta(Z_i, Z_i^*)\right)$$
$$\times \left[\sum_{(\sigma)}\left(\prod_{(i,j)}\prod_{(a_1,a_2)}(i\gamma_a\partial_a)^{-1}_N(Z_i, Z_j)\right)\right]$$
$$\times \exp\left(-\sum_{(i\neq j)}^{N} K^{(\varepsilon)}(Z_i, Z_j, 2\delta(Z_i, Z_i^*))\right), \tag{5.11}$$

where $\sum_{(\sigma)}$ means that we have to sum over all ways of pairing the fermion fields in Eq. (5.10) and $(i\gamma_a\partial_a)^{-1}_N(Z_i, Z_j)$ denotes the flat Dirac Green's function with the Newmann condition along the real axis.

It is instructive to point out the $O(N)$ Chan-Paton factor $Tr\sum_{(\sigma)}(\lambda_{R_{\sigma(1)}} \cdots \lambda_{R_{\sigma(N)}})$ in the expression quoted above. Additionally N should be an even number, thus, implying that the $R^D \times SO(N)$ $K = 2$ bosonic string model possesses a quantum number which is subject to conservation.

Collecting the results of Eqs. (5.6) and (5.11) and evaluating the Faddeev-Popov determinant, we get the final expression conformally regularized for the scalar N-point off-shell scattering amplitudes, Eq. (5.5):

$$\widehat{A}_{R_1} \cdots R_N(p_1, \ldots, p_N) = \left[Tr \sum_{(\sigma)} (\lambda_{R_{\sigma(1)}} \cdots \lambda_{R_{\sigma(N)}}) \right]$$

$$\times \left(\int \mathcal{D}[\delta(Z, Z^*)] \exp\{-S_{\text{eff}}[\delta(Z, Z^*)]\} \right.$$

$$\times \left\{ \int_{R_+^2} \prod_{j=1}^{N} d^2\xi_j^{(H)} \exp\left(\sum_{i=j}^{N} \delta(Z_i, Z_i^*) \right) \right.$$

$$\times \exp\left[-\left(\sum_{i=j}^{N} (p_i^{(A)}; p_j^{(A)}) K^{(\varepsilon)}(Z_i, Z_i, 2\delta(Z_i, Z_i^*)) \right) \right]$$

$$\times \exp\left[-\left(\sum_{i \neq j}^{N} [(p_i^{(A)}; p_j^{(A)}) + 1] K^{(\varepsilon)}(Z_i, Z_j, 2\delta(Z_i, Z_i^*)) \right) \right]$$

$$\left. \left. \times \left[\sum_{(\sigma)} \left(\prod_{(i,j)}^{N} \prod_{(a_1, a_2)} (i\gamma_a \partial_a)^{-1}_{(N)}(Z_{\sigma(i)}; Z_{\sigma(j)}) \right) \right] \right\} \right), \tag{5.12}$$

where the effective Liouville action in the upper half-plane R_+^2 is given by

$$S_{\text{eff}}[\delta(Z, Z^*)] = \frac{26 - D - (N-1)}{12\pi} \int_{R_+^2} d^2\xi [\frac{1}{2}(\partial_a \delta)^2 + \mu^2 e^{2\delta}](\xi)$$

$$\lambda \int_{-\infty}^{+\infty} d\xi_1 e^{\delta(\xi_1, 0)} - D/8\pi \int_{-\infty}^{+\infty} d\xi_1 (\partial_N \delta)(\xi_1, 0) \quad (Z = \xi_1 + i\xi_2). \tag{5.13}$$

Since the solution of the Liouville field theory in R_+^2 was not found yet, which would provide the complete solution of Eq. (5.12), we implement a saddle-point approximation to evaluate Eq. (5.12) as introduced in Ref. 10. By requiring the usual "boundary" restriction $\lambda = -(5/\sqrt{2})\mu$ [cf. Eq. (3.11) of Ref. 10] the (natural) $D \to -\infty$ solution of the theory described by Eq. (5.13) is the Poincar metric in R_+^2: namely,

$$\delta(\xi_1, \xi_2) = \frac{1}{2} \ln \left[\frac{2}{\mu^2 |Z - Z^*|^2} \right] \quad (Z = \xi_1 + i\xi_2). \tag{5.14}$$

We remark that this solution renders the action singular. However the action $S_{\text{eff}}[\delta(Z, Z^*)]$ does not depend on the points Z_i, so it cancels with the normalization factor of the quantum average.

By substituting Eq. (5.14) in Eq. (5.12), we finally get the $D \to -\infty$ (or equivalently $N \to -\infty$) scalar dual Green's function of the $R^D \times SO(N)$ $K = 2$ string model:

$$\begin{aligned}
\widehat{A}_{R_1} \cdots R_N(p_1, \dots, P_N) = [\bar{g}(\varepsilon, p_i)]^N & \left(Tr \sum_{(\sigma)} (\lambda_{R_{\sigma(1)}} \cdots \lambda_{R_{\sigma(N)}}) \right) \\
& \times \int_{R_+^2} \prod_{j=1}^{N} d^2 \xi_j^{(N)} \left(\prod_{j=1}^{N} |Z_i - Z_i^*|^{+\frac{p_i^2}{\pi} - 1} \right) \\
& \times \left(\prod_{i<j} (|Z_i - Z_j'| \, |Z_i - Z_j'^*|)^{[(p_i^{(A)}; p_j^{(A)}) + 1]/\pi} \right) \\
& \times \left(\sum_{(\sigma)} \sum_{\substack{(i,j) \\ (i \neq j)}} \prod_{(a_1, a_2)} [(i\gamma_a \partial_a)_N^{-1} (Z_{\sigma(i)}, Z_{\sigma(j)})]_{a_1 a_2} \right),
\end{aligned} \tag{5.15}$$

where the conformally regularized *dual coupling constant model* is given explicitly by

$$\bar{g}(\varepsilon, p_i) = \left(\frac{2}{\mu^2} \right)^{1/2 - (\sum_{i=j}^{N} p_i^2 / 2\pi N)} \varepsilon^{\sum_{i=j}^{N} p_i^2 / 4\pi N}. \tag{5.16}$$

At this point it is instructive to compare with the similar result obtained for the fermionic string [cf. Eq. (18) of Ref. 11].

In order to determine the associated spectrum we have to find the poles in the external momentum variables $p_i^2) = (p_i; p_i)$. Such poles occur when Z_i and Z_i^* come close together. As a result, (Euclidean) poles exist when

$$p_i^2/\pi = 1 = -1, -2, \dots \quad \text{or} \quad p_i^2/\pi = 0, -1, -2, \dots. \tag{5.17}$$

This fact implies that the spectrum of the $R^D \times SO(N)$ $K = 2$ bosonic string at the $D \to -\infty$ saddle point *is not* sensitive to the axial anomaly and coincides with the spectrum of the pure $D \to -\infty$ fermionic string, a string without tachyans. As another comment, we remark that the absence of the tachyon excitation of these models is entirely due to the covariant treatment of the fermionic degrees which produces the factor $\exp[-\sum_{i=1}^{N} \delta(Z_i, Z_i^*)]$ in Eq. (11), thus, suppressing the tachyon excitation from the pure bosonic spectrum [cf. Ref. 10, Eq. (9.11)]. Opposite to the usual "add-hoc" spectrum projection onto an even "G-parity" sector implemented at the string critical dimension D_c as an operator quantization analysis.

5.3 References

[1] Peter G.O. Freund, Phys. Lett. **151B**, 387 (1985).
[2] Goddard, Nucl. Phys. **B116**, 157 (1976).
[3] C. Lovelace, Phys. lett. **135B**, 75 (1984).
[4] J. Maharana and G. Veneziano, Phys. Lett. **169B**, 177 (1986).
[5] M.B. Green and J. Schwarz, Phys. Lett. **149B**, 117 (1984); **151B**, 21 (1985).
[6] D. Gross, J. Harvey, E. Martinec, and R. Rohm, Nucl. Phys. **B256**, 253 (1985).
[7] P. Candelas, G. Horowitz, A. Strominger, and E. Witten, Nucl. Phys. **B258**, 46 (1985).
[8] A.N. Redlich and Howard J. Schnitzer, Phys. Lett. **167B**, 315 (1986).
[9] D Nemeschansky and S. Yankielowicz, Phys. Rev. Lett. **54**, 620 (1985); **54**, 1736(E) (1985).
[10] B. Durhuus, H.B. Nielsen, P. Olesen, and J.L. Petersen, Nucl. Phys. **B196**, 498 (1982).
[11] Luiz C.L. Botelho, Phys. Lett. **152B**, 358 (1985).
[12] E.S. Fradkin and A.A. Tseytlin, Phys. Lett. **158B**, 316 (1985).
[13] B. Durhuus, P. Olesen, and J.L. Petersen, Nucl. Phys. **B198**, 159 (1982).
[14] S.P. de Alwis, Phys. Lett. **168B**, 59 (1986).
[15] R.E. Gamboa Saravi, F.A. Schaposnik, and H. Vucetich, Phys. Rev. **D30**, 363 (1984).
[16] F. Gliozzi, J. Scherk, and D.J. Olive, Nucl. Phys. **B122**, 253 (1977).

Chapter 6

The Electric Charge Confining in Abelian Rank two Tensor Field Model

We present a Wilson Loop evaluation of the binding energy of electric static charges in an Abelian Rank two Tensor Field Model in R^4 through path integrals.

6.1 Introduction

It has been argued by A.M. Polyakov [1] that the presence of monopoles excitations in the vacuum of Compact Quantum Electrodynamics is the main fact responsible for the confinement or screening of the electric charge of the electron excitations.

In this chapter we intend to show a similar, however diferent dynamics of confinement of electrical charges in the presence of tensor fields by path integral manipulations with abelian Wilson loops path integrals in the framework of dimensional regularization and by phenomenologically representing the dynamics of the theory's vacuum through an effective dynamics of a low-energy rank-two antisymmetric tensor field ([2]) (see appendix for some comments).

6.2 The interquark potential evaluation

Let us thus start our analyzes by recalling a propose to represent quantum field dynamics for a rank-two antisymmetric tensor by means of a Gaussian path integral ([2])

$$
\begin{aligned}
Z = & \int D^F[B_{\mu\nu}] \exp\Big\{ -\frac{1}{4e^2} \int d^4x \Big[B^2_{\mu\nu} \\
& + dB \text{ arsen}(\frac{dB}{m^2}) + m^2 \sqrt{1-(\frac{dB}{m^2})^2} \Big](x) \Big\} \\
& \underset{m^2\to\infty}{\overset{\text{low-energy}}{\sim}} \int D^F[B_{\mu\nu}] \exp\Big\{ -\frac{1}{4e^2} \int d^4x (B_{\mu\nu}[-(\partial^2)+m^2]B_{\mu\nu})(x) \Big\}
\end{aligned} \tag{6.1}
$$

where m^2 is a dimensional transmutation (very heavier) mass parameter which is (phenomenologically) signaling the energy scale where theory's non-perturbative effects are expected to be relevant.

In order to consider the presence of an external electromagnetic field dynamics we consider an interaction with the electromagnetic field by means of a kind of topological term, namely

$$
\begin{aligned}
Z = & \int D^F[B_{\mu\nu}] D^F[A_\mu] \delta(\partial_\mu A_\mu) \exp\Big(-\frac{1}{4e^2} \int d^4x F^2_{\mu\nu}(A)(x) \Big) \\
& \times \exp\Big(-\frac{1}{4e^2} \int d^4x (B_{\mu\nu}[-(\partial^2)+m^2]B_{\mu\nu})(x) \Big) \\
& \times \exp\Big(\theta \int d^4x (B_{\mu\nu}\, {}^*F_{\mu\nu}(A))(x) \Big).
\end{aligned} \tag{6.2}
$$

Let us show the electric charge confinement in the effective theory as described by the path integral eq. (6.2) under the constraint of the fine tunning value $\theta = |m|$.

In order to analyze such phenomenon, we consider the binding energy between two probing electric static charges with opposite charge's signal with a space-time trajectory description $C_{(R,T)}$, the loop boundary of the rectangle $S_{R,T)} = \Big\{ -\frac{T}{2} \le x^0 \le +\frac{T}{2} \,;\, -\frac{R}{2} \le x^1 \le +\frac{R}{2} \Big\}$.

We have thus the standard result for this binding energy in terms of the gauge invariant Wilson loop

$$
E_{\text{bin}}(R) = \lim_{T\to\infty} \left\{ -\frac{1}{T} lg \, \overbrace{\langle \exp(i \int_{C_{(R,T)}} A_\mu(X^\alpha(s) dX^\mu(s)) \rangle}^{W[C_{(R,T)}]} \right\} \tag{6.3}
$$

where the euclidean path integral average, normalized to unity, is defined by eq. (6.2). After realizing the $B_{\mu\nu}(x)$ Gaussian path integral, we arrive at the effective $U(1)$-invariant path integral result for the Wilson Loop observable inside eq. (6.3).

$$\langle W[C_{(R,T)}]\rangle = \int D^F[A_\mu]\delta^{(F)}(\partial_\mu A_\mu)$$
$$\times \exp\Big\{-\frac{1}{2e^2}\int d^4x d^4y A_\mu(x)[(-\partial^2)_x\delta^{(4)}(x-y)$$
$$-\theta^2(-\partial^2)_x(-\partial_x^2+m^2)^{-1}(x,y)]A^\mu(y)\Big\}$$
$$\times \exp\left(i\int_{C_{(R,T)}} A_\mu(X^\alpha(s)dX^\mu(s)\right). \quad (6.4)$$

The searched binding energy takes thus the simple form in a dimensional regularized integral-distributional form

$$E_{\rm bin}(R) = \lim_{T\to\infty}\left\{-\frac{e^2}{T}\int\frac{d^\nu k}{(2\pi)^\nu}f_\mu(k,C_{(R,T)})D(k^2)f^\mu(-k,C_{(R,T)})\right\} \quad (6.5)$$

with the rectangle form factors ([4])

$$f_\mu(k,C_{(R,T)}) = \oint_{C_{(R,T)}} e^{-i(k_0x_0(s)+k_1x_1(s))}\frac{dx_\mu(s)}{ds}$$
$$= \iint_{S_{(R,T)}} \varepsilon^{\mu\nu}\left[\frac{\partial}{\partial x_\nu}(e^{-ik_\alpha x_\alpha})\right]d^2x_\alpha$$
$$= \begin{cases} -\frac{4}{k_0}\sin(\frac{k_0T}{2})\sin(\frac{k_1R}{2}) & \text{for } \mu=0 \\ +\frac{4}{k_1}\sin(\frac{k_0T}{2})\sin(\frac{k_1R}{2}) & \text{for } \mu=1 \end{cases} \quad (6.6)$$

The effective electromagnetic propagator, local in momentum space, is explicitly given by

$$D(k^2) = \left[k^2 - \frac{\theta^2k^2}{k^2+m^2}\right]^{-1} = \frac{k^2+m^2}{k^4+k^2(m^2-\theta^2)}. \quad (6.7)$$

Now one can easily see that for the fine tunning choice $m=+\theta$, one obtains after straightforwardly calculations ([4]), the expected confining Cornell form ([5]) for the binding energy

$$E_{\rm bind}(R) = \left(-\frac{e^2}{4\pi R}\right) + ((4\pi e^2m^2)R). \quad (6.8)$$

It is worth that for $m\neq\theta$, or $\theta = i\overline{\theta}$ ($\overline{\theta}\in R$), call the reader attention that the binding energy is of the Yukawa form and thus leading to a screening picture for the inter-electronic potential for compact QED in R^4.

At this point, let us comment and compare ours results with the attempts done in ref.[3] through Hamiltonian methods for a similar quantum field model.

We feel that these works may be not well-defined by the somewhat cumbersome use of cut-offs (otherwise infinite) ordinary integrals. For instance, in these refs.[5], it has been obtained the following expression for the binding energy (see eqs. (17)-(29) of ref.[3] where $R = (\vec{y} - \vec{y}')$, $M^2 = m^2 + e^2$ in the refs author's notation):

$$E_{\text{bind}}(R) = V^{(1)}_{(R)} + V^{(2)}_{(R)} \tag{6.9}$$

with

$$\begin{aligned} V^{(1)}_{(R)} &= -\frac{e^2}{2} \int d^3x \int_{\vec{y}}^{\vec{y}'} dz'_i \, \delta^{(3)}(x - z') \left(\frac{1}{\nabla_x^2 - M^2} \nabla_x^2 \right) \left(\int_{\vec{y}}^{\vec{y}'} dz^i \, \delta^{(3)}(x - z) \right) \\ &= (e^2 \, e^{-MR})/4\pi R \end{aligned} \tag{6.10}$$

and

$$\begin{aligned} V^{(2)}(R) &= +\frac{e^2 m^2}{2} \int d^3x \int_{\vec{y}}^{\vec{y}'} dz'_i \, \delta^{(3)}(x - z') \left(\frac{1}{\nabla_x^2 - M^2} \right) \int_{\vec{y}}^{\vec{y}'} dz^i \, \delta^{(3)}(x - z) \\ &= -\frac{e^2}{4\pi} (e^{-MR}/R) + \frac{e^2 M}{8\pi} R \, \ell n \left(1 - \frac{\varepsilon^3}{M^2} \right) \end{aligned} \tag{6.11}$$

where ε is a cutt-off to be imposed in order to make sense for their divergent ordinary integrals. However, it appears that this process of handling distributions has drawbacks, since one can easily use the screening result eq. (6.10) to evaluate eq. (6.11) through the use of the simple operatorial decomposition

$$\begin{aligned} V^{(2)}(R) = \frac{e^2 m^2}{2} & \int d^3x \int_{\vec{y}}^{\vec{y}'} dz'_i \, \delta^{(3)}(x - z') \left(\frac{(\nabla_x)^2}{(\nabla_x)^2 (\nabla_x^2 - M^2)} \right) \\ & \times \left(\int_{\vec{y}}^{\vec{y}'} dz^i \, \delta^{(3)}(x - z) \right) \\ = \frac{e^2 m^2}{2} \Bigg\{ & \int d^3x \int_{\vec{y}}^{\vec{y}'} dz_1^i \, \delta^{(3)}(x - z') \left[-\frac{1}{M^2} \frac{1}{(\nabla_x)^2} + \frac{1}{M^2} \frac{1}{(\nabla_x^2 - M^2)} \right] \\ & \times \left(\int_{\vec{y}}^{\vec{y}'} dz^i \, \delta^{(3)}(x - z) \right) = -\frac{e^2 M^2}{2M^2} \left(\frac{1}{R} \right) + \frac{e^2 m^2}{2M^2} \left(\frac{e^{-MR}}{4\pi R} \right) \end{aligned} \tag{6.12}$$

which clearly is finite and differs from the their claimed cut-off implicit confining result. At this point let us suggest that the use of Hamiltonian formalism to handle mathematically gauge theories, specially Yang-Mills theory and its variants is somewhat difficulted by the fact that one must fix the gauge and this makes recourse for formal BRST realizations of Gauge Invariance in the observables evaluation in order to make the results physically acceptable. Certainly such somewhat formal technique making use of "infinitesimal" grassmanian parameters in a theory with solely physical bosons fields as imputs must be applied very carefully outside the well-founded evaluations of matrix (on-sheel) scattering amplitudes which does not exist in pure Yang-Mills theory due to severe infrared divergencies. That was the main reason of reformulating all the gauge theories by means of gauge invariant path integrals and mainly with the objective that in the confining phase all the evaluations should be done non-perturbativelly (see appendix for supplementary comments).

6.3 Appendix A: The dynamics of the $QCD(SU(\infty))$ tensor fields from strings

In this appendix we intend to high-light on some ideas and loop space formulae about non-space time supersymmetric random surfaces representations for bosomic $QCD(SU(N_c))$ (including the t'Hooft planer diagrams limit of $N_c \to \infty$).

Alexandre M. Polyakov in the article ref [6], has argued that one possible random surface path integral representation for Wilson Loops in bosonic $QCD(SU(N_c))$ shall be given by our discovery that mathematical bosonic strings interacting with A Migdal intrinsic fermionic Strings degree of freedom in the $SU(3) \times SU(2) \times U(1)$ fundamental representation [in an non-abelian bosonized form] and interacting with a rank two antisymmetric tensor field through the closed string world-sheet orientation tangent tensor namely ([6], [7]).

$$W[C] = \Big\langle \sum_{S_i \partial s = \ell} \sum_{\{g\}} \Big\{ \exp\Big[- \int d^2\xi\, \zeta^{\mu\nu}(X^\beta(\xi)).\, \mathrm{Tr}(g^{-1} B_{\mu\nu}(X(\xi)g) \Big]$$

$$\exp[-(\sigma - \text{model action for } g \text{ with Wess-Zumino-Novikov}$$

$$\text{Topological Term})] \Big\} \Big\rangle_B \tag{A-1}$$

The author of the second set of ref.[7] present arguments that the average over the $B_{\mu\nu}(x)$, probably representing the non-trivial random flux structure of the QCD-vacuum should be defined by a pure white-noise set of random fluxes. In the path integral language ([6],]8]), the B-average should be defined by the functional measure:

$$\Big\langle\ \Big\rangle_B = \int D^F[B_{\mu\nu}(x)] \exp\left\{-\frac{1}{4(g_\infty)^2}\int d^4x\, B^2_{\mu\nu}(x)\right\} \tag{A-2}$$

where $g^2_\infty = \lim_{N_i\to\infty}(g^2N_c) < \infty$ denotes the $N_c = +\infty$ t'Hooft QCD coupling constant.

It is very important to remark now that eq. (A-1) with the White-Noise random flux average eq. (A-2), satisfies the QCD loop wave equation exactly under the geometrical hypothesis that the bosonic surfaces (string world-sheets) defining the QCD string should satisfies the self-intersecting orthogonality tangent plane constraints $(\xi = (s,\sigma)), X^\mu(s,0^+) = \ell^u(s))$

$$\begin{aligned}
&\zeta_{\mu\nu}(X^\beta(s,\sigma)\cdot\zeta^{\mu\nu}(X^\beta(s',\sigma')) = 0 \quad \text{if} \quad s\neq s' \\
&\zeta^{\mu\nu}(X^\beta(s,\sigma))\,\zeta_{\mu\nu}(X^\beta(s',\sigma)) \neq 0 \quad \text{if} \quad s = s'
\end{aligned} \tag{A-3}$$

Note that the second constraint in eq. (A-3) means that one allows non-trivial topology in the intrinsic (mathematical) string time-direction evolution $0\leq\sigma<\infty$. However, the first geometrical constraint may be connected (not proved yet) to the fact that the bosonic loops $X^\mu(s,\bar\sigma)$ (with $\bar\sigma$ fixed) should physically corresponding to R^4 space-time euclidean quark-antiquark trajectories on the fermionic quark functional determinant in the presence of an abelian color singlet vectorial source, for instance:

$$\begin{aligned}
&\ell g\ \det(\partial\!\!\!/(A_\mu)+J_\mu) = -\frac{1}{2}\int_0^\infty \frac{d\Gamma}{T}\left\{\int d^4\,z^\beta\int_{X^\beta(0)=X^\beta(T)=z^\beta} D^F[X(s)]\right. \\
&\exp\left(-\frac{1}{2}\int_0^T ds\cdot X^2(s)\right)\times\exp\left(ie\int_0^T ds\,J_\mu(X(s))\right) \\
&\mathrm{Tr}\left[\mathbb{P}\ \mathrm{Spin}\ \left(\exp\frac{g^2}{4i}\int_0^T ds[\gamma^\mu,\gamma^\nu]\left[\frac{\delta}{\delta\sigma_{\mu\nu}(X(s))}\right]\right)\right] \\
&\left.\times\frac{1}{N_c}\mathrm{Tr}_{\mathrm{color}}\,\mathbb{P}\left(\left[\exp ig\int_0^T A_\mu(X(s))dX^\mu(s)\right)\right]\right\}
\end{aligned} \tag{A-4}$$

with $\frac{\delta}{\delta\sigma_{\mu\nu}(x^\beta)}$ denoting the formal Mandelstam-Migdal area functional derivative $\frac{\delta}{\delta\sigma_{\mu\nu}(X^\beta(s))} = \frac{\delta}{\delta[(\dot X_\mu X_\nu - \dot X_\nu X_\mu)(s)]}$ and satisfying thus the Pauli Exclusion principle which by its turn translates geometrically by the condition that at non-trivial proper-time trajectories self-intersections

$X^\mu(s) = X^\mu(s')$ $(0 < s, s' < T)$, we have always that $\dfrac{dX^\mu(s)}{ds} \cdot \dfrac{dX_\mu(s')}{ds'} \equiv 0$. This important Pauli Exclusion random surfaces probably signals that strings representations for scalar quark QCD is ill-defined in relation to strings representations for fermionic quark QCD as proposed in refs [7], [8].

At this point one may expect that in the very low energy scale of QCD, one can integrate out all the strings degrees of freedom and arrives at the somewhat crude however local effective rank-two antisymetric tensor field simplest action eq. (A-1), but added with Abelian Wilson Loop as a sort of remnants of the full non-Abelian colour Wilson Loop phase factor. Note that the closed loops are already under a fixed proper-time parametrization, as one can see from the loop space expression for eq. (A-4).

As an important point to be singlet out is that the Loop Space/String program for QCD semms much better defined if one introduces supersymetry on the quark-antiquark world lines and string (spinning) world sheets to handle directly the Lorentz spin content of the quarks, instead of the obligatory operational loop Feynman Lorentz spin factor inside eq. (A-1) with the operational expression:

$$\Phi^{\text{Spin}}[\ell] = \mathbb{P}_{\text{Dirac}} \left\{ \exp\left(-\frac{g_\infty^2}{2} \oint_C ds [\gamma^\mu, \gamma^\nu] \times \left[\frac{\delta}{\delta\zeta_{\mu\nu}(X^\beta(s,\sigma))} \right]_{\sigma \to 0^+} \right) \right\} \tag{A-5}$$

equivalently written by the introduction of a new set of intrinsic neutral Fermion fields in the string world sheet: $\{\psi^\mu(\xi)\}_{\mu=0,1,2,3}$ and representing the "String Lorentz Spin". Namely.

$$\begin{aligned} \Phi^{\text{Spin}}_{\alpha\beta}[\ell] &= \int [D^F \psi^\mu_{(\xi)}] \times \exp\left[-\frac{1}{2} \int d^2\xi (\psi^\mu (i\gamma) \psi_\mu)(\xi) \right] (\psi_\alpha(0,0) \psi_\beta(T,0)) \\ &\times \exp\Bigg(-\frac{g_\infty^2}{2} \int d^2\xi (\psi_\alpha [\gamma^\mu, \gamma^\nu]_{\alpha\beta} \psi_\beta)(\xi) \\ &\times \text{Tr}\left[g^{-1}(\xi) \frac{\delta}{\delta\zeta_{\mu\nu}(X^\beta(\xi))} g(\xi) \right) \Bigg) \end{aligned} \tag{A-6}$$

It is hoped that some sort of solitons/Lorentz Spin Vertexs in the QCD Spinning, flavor charged string theory eq. (A-1)–eq. (A-5) should be candidates to describe the Baryons Physics, basic issue not yet handled in the loop space framework and others holographic proposals for strings representations in supersymmetric Yang-Mills theory.

6.4 References

[1] A.M. Polyakov, Gauge Field and Strings. Harwood Academic Chur., Switzerland, 1987.
[2] A.M. Polyakov, Particle Phys. B486, 23, (1997).
[3] P. Gaete, C. Wotzasik, Phys. Lett. B 601, 108, (2004).
[4] Luiz C.L. Botelho, Phys. Rev. 700, 045010, (2004).
[5] Eichten, Gottpried, T.Kinoshita, K.O. Lane, T.M. Yau, Phys. Rev 170, 3050, (1978).
[6] Alexandre M. Polyakov, String Theory and Quark Confinement – arKiv: hep-th/971100201, 1 Nov 1997.
[7] Luiz C.L. Botelho - CBPF - NF - 051/85 (Pre-Print) - (1985)
- CBPF - NF - 034/86 (Pre-Print) - (1986)
- CBPF - NF - 045/86 (Pre-Print) - (1986)
- Brazilian Journal of Physics, vol. 21, nº 4, (1991)
- Brazilian Journal of Physics, vol. 19, nº 3, (1989)
- Caltech preprint 68 - 1444, (1987)
[8] Luiz C.L. Botelho, J. Math. Phys. 30 (9), 2160, (1989).
[9] A.M. Polyakov, Nucl. Phys. B 486, 28, (1997)
[10] L.C.L. Botelho, Phys. Lett. B 152, 358, (1985)
- Phys. Lett. B 169, 428, (1986)

6.5 Appendix B: Path-integral bosonization for a nonrenormalizable axial four-dimensional fermion model

We study the bosonization and exact solubility of a nonrenormalizable four-dimensional axial fermion model in the framework of anomalous chiral path integrals.

6.6 Introduction

The study of two-dimensional fermion models in the framework of chiral anomalous path integrals has been shown to be a powerful nonperturbative technique to analyze the two-dimensional bosonization phenomenon.

It is the purpose of this appendix to implement this nonperturbative technique to solve exactly a nontrivial and nonrenormalizable four-dimensional axial fermion model which generalizes for four dimensions the two-dimensional model studied in Ref. 2.

6.7 The model

Let us start our analysis by considering the (Euclidean) Lagrangean of the proposed Abelian axial (mathematical) model

$$\mathcal{L}_1(\psi, \bar{\psi}, \phi) = \bar{\psi}\gamma_\mu(i\partial_\mu - g\gamma_5\partial_\mu\phi)\psi + \frac{1}{2}g^2(\partial_\mu\phi)^2 + V(\phi), \qquad \text{(B-1)}$$

where $\psi(x)$ denotes a massless four-dimensional fermion field, $\phi(x)$ a pseudoscalar field interacting with the fermion field through a pseudoscalar derivative interaction, and $V(\phi)$ is a ϕ self-interaction potential given by

$$V(\phi) = \frac{-g^4}{12\pi^4}\phi(\partial_\mu\phi)^2(-\partial^2\phi) + \frac{g^2}{4\pi^2}(-\partial^2\phi)(-\partial^2\phi). \qquad \text{(B-2)}$$

the presence of the above ϕ potential is necessary to afford the exact solubility of the model as we will show later [eq. (18)].

The Hermitian γ matrices we are using satisfy the (Euclidean) relations

$$\{\gamma_\mu, \gamma_\nu\} = 2\delta_{\mu\nu}, \quad \gamma_5 = \gamma_0\gamma_1\gamma_2\gamma_3. \qquad \text{(B-3)}$$

The Lagrangian $\mathcal{L}_1(\psi, \bar{\psi}, \phi)$ is invariant under the global Abelian and chiral Abelian groups

$$\psi \to e^{i\alpha}\psi, \quad \psi \to e^{i\gamma_5\beta}\psi, \quad (\alpha, \beta) \in \mathbb{R}, \qquad \text{(B-4)}$$

with the Noether conserved currents at the classical level:

$$\partial_\mu(\bar{\psi}\gamma^5\gamma^\mu\psi) = 0, \quad \partial_\mu(\bar{\psi}\gamma_\mu\psi) = 0. \qquad \text{(B-5)}$$

In the framework of path integrals, the generating functional of the correlation functions of the mathematical model associated with the Lagrangian $\mathcal{L}_1(\psi, \bar{\psi}, \phi)$ is given by

$$\begin{aligned} Z[J, \eta, \bar{\eta}] = \frac{1}{Z[0,0,0]} \times \int D[\phi]D[\psi]D[\bar{\psi}] \\ \times \exp\left(-\int d^4x[\mathcal{L}_1(\psi, \bar{\psi}, \phi) + J\phi + \bar{\eta}\psi + \bar{\psi}\eta](x)\right). \end{aligned} \qquad \text{(B-6)}$$

In order to generalize for four dimensions the chiral anomalous path itnegral bosonization technique as in Refs. [2] and [3], we first rewrite the full Dirac operator in the following suitable form:

$$\begin{aligned} \not{D}[\phi] &= i\gamma_\mu(\partial_\mu + ig\gamma_5\partial_\mu\phi) \\ &= \exp(ig\gamma_5\phi)(i\gamma_\mu\partial_\mu)\exp(ig\gamma_5\phi). \end{aligned} \qquad \text{(B-7)}$$

Now we proceed as in the two-dimensional case by decoupling the fermion field from the pseudoscalar field $\Phi(x)$ in the Lagrangian $\mathcal{L}_1(\psi, \bar{\psi}, \phi)$ by making the chiral change of variables:

$$\begin{aligned} \psi(x) &= \exp[ig\gamma_5\phi(x)]\chi(x), \\ \bar{\psi}(x) &= \bar{\chi}(x)\exp[ig\gamma_5\phi(x)]. \end{aligned} \tag{B-8}$$

On the other hand, the fermion measure $D[\psi]D[\bar{\psi}]$, defined by the eigenvectors of the Dirac operator $\not{D}[\phi]$, is not invariant under the chiral change and yields a nontrivial Jacobian, as we can see from the relationship

$$\begin{aligned} &\int D[\psi]D[\bar{\psi}]\exp\left[-\int d^4x(\bar{\psi}\,\not{D}[\psi]\psi)(x)\right] \\ &\quad = \det(\not{D}[\phi]) \\ &\quad = J[\phi]\int D[\chi]D[\bar{\chi}]\exp\left(-\int |d^4x(\bar{\chi}i\gamma_\mu\partial_\mu\chi)(x)\right). \end{aligned} \tag{B-9}$$

Here $J[\phi] = \det(\not{D}[\phi])/\det(\not{D}[\phi = 0])$ is the explicit expression for this Jacobian.

It is instructive to point out that the model displays the appearance of the axial anomaly as a consequence of the nontriviality of $J[\phi]$, i.e., $\partial_\mu(\bar{\psi}\gamma_\mu\gamma_5\psi)(x) = \{(\delta/\delta\phi)J[\phi]\}(x)$.

So, to arrive at a complete bosonization of the model Eq. (B-6) we face the problem of evaluating $J[\phi]$.

Let us, thus, compute the four-dimensional fermion determinant $\det(\not{D}[\phi])$ exactly, In order to evaluate it, we introduce a one-parameter family of Dirac operators interpolating the free Dirac operator and the interacting one $\not{D}^{(\zeta)}[\phi]$: namely,

$$\not{D}^{(\zeta)}[\phi] = \exp(ig\gamma_5\zeta\phi)(i\gamma_\mu\partial_\mu)\exp(ig\gamma_5\zeta\phi) \tag{B-10}$$

with $\zeta \in [0, 1]$.

At this point we introduce the Hermitian continuation of the operator $\not{D}^{(\zeta)}[\phi]$ by making the analytic extension in the coupling constant $\bar{g} = ig$. This procedure has to be done in order to define the functional determinant by the proper-time method since only in this way $(\not{D}^\zeta[\phi])$ can be considered as a (positive) Hamiltonian.

The justification for this analytic extension in the model coupling constant is due to the fact that typical interaction energy densities such as $\bar{\psi}\gamma^5\psi, \bar{\psi}\gamma_\mu\Delta^\mu\psi$, which are real in Minkowski space-times, become complex after continuation in Euclidean space-time. As a consequence, the above analytic coupling extension must be done in the proper-time regularization for the Dirac functional determinant.

By using the property

$$\frac{d}{d_\zeta} \not{D}^\zeta[\phi] = \bar{g}\gamma_5\phi \not{D}^\zeta[\phi] + \not{D}^\zeta[\phi]\gamma_5\bar{g}\phi, \tag{B-11}$$

we can write the following differential equation for the functional determinant (see Ref. [4] and chapter 3):

$$\frac{d}{d_\zeta} \ln\det \not{D}^\zeta[\phi]$$
$$= 2\lim_{\varepsilon\to 0^+} \int d^4x \, \mathrm{Tr}\langle x|(\bar{g}\gamma_5\phi)\exp\{-\varepsilon(\not{D}^\zeta[\phi])^2\}|x\rangle, \tag{B-12}$$

where Tr denotes the trace over Dirac indices.

The diagonal part of $\exp\{-\varepsilon(\not{D}^\zeta[\phi])^2\}$ has the asymptotic expansion

$$\langle x|\exp\{-\varepsilon(\not{D}^\zeta[\phi])^2\}|x\rangle$$
$$\underset{\varepsilon\to 0^+}{\sim} \frac{1}{16\pi^2\varepsilon^2}\left(1 + \varepsilon H_1(\phi) + \frac{\varepsilon^2}{2!}H_2(\phi)\right) \tag{B-13}$$

with the Seeley coefficients given by (see the Appendix)

$$H_1(\phi) = \zeta\bar{g}\gamma_5\partial^2\phi - \bar{g}\zeta^2(\partial_\mu\phi)^2 \tag{B-14}$$

and

$$H_2(\phi) = 2\partial^2 H_1(\phi) + 2\zeta\bar{g}\gamma_5(\partial_\mu\phi)\partial_\mu H_1(\phi) + 2[H_1(\phi)]^2. \tag{B-15}$$

By substituting Eqs. (B-13)–(B-15) into Eq. (B-12), we obtain finally the result for the above-mentioned Jacobian:

$$J[\psi] = J_0[\phi,\varepsilon]J_1[\phi]. \tag{B-16}$$

Here $J_0[\phi,\varepsilon]$ is the ultraviolet cutoff-dependent Jacobian term

$$J_0[\phi,\varepsilon] = \exp\left(\frac{g^2}{4\pi^2\varepsilon}\int d^4x[\phi(-\partial^2)\phi](x)\right) \tag{B-17}$$

and $J_1(\phi)$ is the associated Jacobian finite part

$$J_1[\phi] = \exp\left(-\frac{g^2}{4\pi^2}\int d^4x(-\partial^2\phi)(-\partial^2\phi)(x)\right)$$
$$\times \exp\left(\frac{g^4}{12\pi^2}\int d^4x[\phi(\partial_\mu\phi)^2(-\partial^2\phi)](x)\right). \tag{B-18}$$

From Eqs. (B-17) and (B-1), e can see that the (bare) coupling constant g^2 gets an additive (ultraviolet) renormalization. Besides the Jacobian term cancels with the chosen potential $V(\phi)$ in Eq. (B-2).

As a consequence of all these results, we have the following expression for $Z[J,\eta,\bar{\eta}]$ with the fermions decoupled:

$$\begin{aligned} Z[J,\eta,\bar{\eta}] = \frac{1}{Z[0,0,0]} \int D[\phi]D[\chi]D[\bar{\chi}] \\ \times \exp\Big(-\int d^4x[\frac{1}{2}\bar{x}(i\gamma_\mu\partial_\mu)\chi + \frac{1}{2}g_R^2(\partial_\mu\phi)^2 \\ + \bar{\eta}\exp(ig_R\gamma_5\phi)\chi + \bar{\chi}\exp(ig_R\gamma_5\phi)\eta + J\phi](x)\Big). \end{aligned} \quad \text{(B-19)}$$

This expression is the main result of our paper and should be compared with the two-dimensional analogous generating functional analyzed in Ref. [2]. Now we can see that the quantum model given by Eq. (B-6) although being nonrenormalizable by usual power counting and Feynman-diagrammatic analysis it still has nontrivial and exactly soluble Green's functions. For instance, the two-point fermion correlation function is easily evaluated and produced the result

$$\begin{aligned} \langle\psi_\alpha(x_1)\bar{\psi}_\beta(x_2)\rangle = S^F_{\alpha\beta}(x_1 - x_2; m = 0) \\ \times \exp[-\Delta_F(x_1 - x_2; m = 0)]. \end{aligned} \quad \text{(B-20)}$$

Here $\Delta(x_1 - x_2; m = 0)$ is the Euclidean Green's function of the massless free scalar propagator. We notice that correlation functions involving fermions $\psi(x), \bar{\psi}(x)$ and the pseudoscalar field $\phi(x)$ are easily computed too (see Ref. [2]).

As a conclusion of our paper let us comment to what extent our proposed mathematical Euclidean axial model describes an operator quantum field theory in Minkowski space-times.

In the operator framework the fields $\psi(x)$ and $\phi(x)$ satify the following wave equations:

$$\begin{aligned} i\gamma_\mu\partial_\mu\psi = ig\gamma_\mu\gamma_5(\partial_\mu\phi)\psi, \\ \phi = \partial_\mu(\bar{\psi}\gamma_\mu\gamma_\mu\psi) + \frac{\delta\Delta}{\delta\phi}. \end{aligned} \quad \text{(B-21)}$$

It is very difficult to solve exactly Eq. (B-21) in a pure operator framework because the model is axial anomalous $[\partial_\mu(\bar{\psi}\gamma_\mu\gamma_5\psi) \neq 0]$. However the path-integral study Eq. (B-19) shows that the operator solution of Eq. (B-21) [in terms of free (normal-ordered) fields] is given by

$$\Psi(x) =: \exp[i\gamma_5\phi(x)] : \chi(x) \quad \text{(B-22)}$$

since it is possible to evaluate exactly the anomalous fivergence of the above-mentioned axial-vector current and, thus, choose a suitable model potential $V(\phi)$ which leads to the above simple solution.

It is instructive to point out that the operator solution Eq. (B-22) coincides with the operator solution of the $U(1)$ vectorial model analyzed by Schroer in Ref. [8] [the only difference between the model's solutions being the γ_5 factor in the phase of Eq. (B-22)].

Consequently we can follow Schroer's analysis to conclude that the Euclidean correlation functions Eq. (B-20) define Wightman functions which are distributions over certain class of analytic test functions. But the model suffers the problem of the nonexistence of time-ordered Green's functions which means that the proposed axial model in Minkowski space-times does not satisfy the Einstein causality principle.

Finally we remark that the proposed model is to a certain extent less trivial than the vectorial model since for nondynamical $\phi(x)$ field the associated S matrix is nontrivial and is given in a regularized form by the result

$$\begin{aligned} S &= T\left[\exp\left(i\int_{-\infty}^{+\infty}(\bar{\psi}\gamma_\mu\gamma_5\partial_\mu\phi\psi)(x)d^4x\right)\right] \\ &= \exp\left[-i\int_{-\infty}^{+\infty}\phi(x)\left(\frac{\delta}{\delta\phi(x)}J_\varepsilon[\phi]\right)(x)d^4x\right] \end{aligned} \tag{B-23}$$

with $J_\varepsilon[\phi]$ expressed by Eq. (B-16).

6.8 Appendix C

We now briefly calculate the asymptotic term of the second-order positive differential elliptic operator:

$$(\not{D}^\zeta[\phi])^2 = (-\partial^2) - (\bar{g}\gamma_5\zeta\partial_\mu\phi)\partial_\mu - \bar{g}\zeta\gamma_5\partial^2\phi + (\partial_\mu\phi)^2. \tag{C-1}$$

For this study, let us consider the more general second-order elliptic four-dimensional differential operator (non-necessarily) Hermitian in relation to the usual normal in $L^2(\mathbb{R}^D)$ (Ref. [7]):

$$\mathcal{L}_x = -(\partial^2)_x + a_\mu(x)(\partial_\mu)_x + V(x). \tag{C-2}$$

Its evolution kernel $K(x, y, \zeta) = \langle x|e^{-\zeta D}|y\rangle$ satisfies the heat-kernel equation

$$\begin{aligned} &\frac{\partial}{\partial\zeta}K(x, y; zeta) = -\mathcal{L}_x K(x, y; \zeta), \\ &\lim_{\zeta\to 0^+} K(x, y; \zeta) = \delta^{(D)}(x - y). \end{aligned} \tag{C-3}$$

The Green's function $K(x,y,\zeta)$ has the asymptotic expansion

$$\lim_{\zeta\to 0^+} K(x,y;\zeta) \sim K_0(x,y;\zeta)\left(\sum_{m=0}^{\infty}\zeta^m H_m(x,y)\right), \tag{C-4}$$

where $K(x,y,\zeta)$ is the evolution kernel for the n-dimensional Laplacian $(-\partial^2)$. By substituting Eq. (C-4) into Eq. (C-3) and taking the coinciding limit $x\to y$, we obtain the following recurrence relation for the coefficients $H_m(x,x)$:

$$\begin{aligned}\sum_{n=0}^{\infty}\frac{1}{n!}\zeta^n H_{n+1}(x,x) = -\Bigg(&\sum_{n=0}^{\infty}\frac{\zeta^n}{n!}(-\partial_x^2)H_n(x,x)\\ &+ a_n(x)\sum_{n=0}^{\infty}\frac{\zeta^n}{n!}\partial_\mu^x H_n(x,x)\\ &+V(x)\sum_{n=0}^{\infty}\frac{\zeta^n}{n!}H_n(x,x)\Bigg).\end{aligned} \tag{C-5}$$

For $D=4$, Eq. (C-5) yields the coefficients

$$\begin{aligned}&H_0(x,x) = \mathbf{1}_{4\times 4},\\ &H_2(x,x) = -V(x),\\ &H_2(x,x) = 2[-\partial_x^2 V(x) + a_\mu(x)\partial_\mu V(x) + V^2(x)].\end{aligned} \tag{C-6}$$

Now substituting the value $a_\mu(x) = [-\zeta\bar{g}\partial_\mu\phi(x)]\mathbf{1}_{4\times 4}$ and $V(x) = [(-\bar{g}\zeta\gamma_5\partial_x^2\phi + \partial_\mu\phi)^2]\mathbf{1}_{4\times 4}$ into Eq. (C-6) we obtain the result Eqs. (B-14) and (B-15) quoted in the main text.

6.9 References

[1] K. Fujikawa, Phys. Rev. **D 21**, 2848 (1980); R. Roskies and F. Schaposnik, *ibid.* **23**, 518 (1981).

[2] L. Botelho, Phys. Rev. **D 31**, 1503 (1985); A. Das and C.R. Hagen, *ibid.* **32**, 2024 (1985).

[3] L. Botelho, Phys. Rev. **D 33**, 1195 (1986).

[4] O. Alvarez, Nucl. Phys. **B 238**, 61 (1984).

[5] V. Romanov and A. Schwartz, Teor. Mat. Fiz. **41**, 190 (1979).

[6] K. Osterwarder and R. Schrader, Helv. Phys. Acta **46**, 277 (1973).

[7] P.B. Gilkey, in *Proceedings of Symposium on Pure mathematics*, Stanford, California, 1973, edited by S.S. Chern and R. Osserman (American Mathematical Society, Providence, RI, 1975), Vol. 127, Pt. II, p. 265.

[8] B. Schroer, J. Math. Phys. **5**, 1361 (1964); *High Energy Physics and Elementary Particles* (IAEC, Vienna, 1965).
[9] K. Bardakci and B. Schroes, J. Math. Phys. **7**, 16 (1966).
[10] A. Jaffe, Phys. Rev. **158**, 1454 (1966).

Chapter 7

Infinities on Quantum Field Theory: A Functional Integral Approach

We analyze on the formalism of probabilities measures-functional integrals on function space, the problem of infinities on Euclidean field theories. We also clarify and generalize ours previous published studies on the subject.

7.1 Introduction

The subject of infinities of Euclidean Quantum Field Theories, or in general Minkowskian Quantum Field Theories, has been well grounded with practical Calculations on Quantum Physics since its full inception on 1950 years ([1]).

The purpose of this chapter is to study the nature of ultraviolet infinities on Euclidean Quantum Field Path integral through an analitically regularized, mathematical rigorously path integrals. This study is presented in Section 1. In Section 2, we clarify our previous studies on the subject ([2], [7]) by analyzing in detail all those estimates leading to a correct understanding of the well-known problem of coupling constant renormalization on QFTs in a two-dimensional scalar interacting field model. In Section 3, mainly of pedagogical purpose, we present a mathematically detailed construction of the Wiener Kac functional measure. In Section 4, we analyze the mathematical structure of the so called Feynman geometrodynamical propagation on Euclidean QFT, by using the mathematical rigorous results of Section 3.

In several appendixes, we add several calculations useful to understand the bulk of this chapter.

7.2 Infinities on Quantum Field Theory on the Functional Integral Formalism

Euclidean quantum fields are expected to be mathematically defined objects by the analytic continuation of Minkowskian self-adjoints quantum fields to imaginary time, where this precise analytic continuation is well expressed through the famous Bargmann-Hall-Wightman theorem ([1]). It is thus expected that Euclidean self-adjoints quantum field operators on R^D should be naturally constructed (from a rigorous mathematical point of view) from a probability measure $d\mu(\phi)$ on the L. Schwartz Tempered Districutions $S'(R^D)$. It is also expected that the following functional, the called theory's generating functional $Z(j) \in C(S(R^D), R)$ should furnishes the bridge of such time analytic extension though the identity

$$\langle \Omega_{\rm vac}^{\rm eucl} | \exp i\phi(j) | \Omega_{\rm vac}^{\rm eucl} \rangle = Z(j). \tag{7.1}$$

Here $\phi(j) = \langle \phi, j \rangle$ is the usual canonical pairing between the distribution valued euclidean field operator ϕ (take from here on as a neutral scalar field for simplicity of our exposition).

Since eq. (7.1) is a Bochner-Martin positive definite functional on $C(S(R^D), R)$ one can apply the Minlos Theorem to represent eq. (7.1) by means of a probability measure on $S'(R^D)$

$$Z(j) = \int_{S'(R^D)} d\mu(\phi) \exp i\phi(j). \tag{7.2}$$

For example, massive free scalar fields on R^D given by the explicity generating functional below (exponential of a continuous bilinear form on $S(R^D)$)

$$Z(j) = \exp\left(-\frac{1}{2} T_{(-\Delta+m^2)^{-1}j}(j) \right). \tag{7.3}$$

Here our T (continuous application) is explicited by the formula

$$T\colon (S(R^D), \text{ strong topology}) \to (S'(R^D), \text{ weak topology})$$
$$j \to T_{(-\Delta+m^2)^{-1}j} \tag{7.3-a}$$

where for $f \in S(R^D)$, we have

$$T_{(-\Delta+m^2)^{-1}j}(f) = \int_{R^D} ((-\Delta + m^2)^{-1}j)(x) f(x) d^D x. \tag{7.3-b}$$

As a consequence of above exposed, one simple application of Minlos-Bochner Theorem to eqs. (7.3-a)/(7.3-b) give us the massive free scalar euclidean generating functional

$$Z(j) = \exp\left(-\frac{1}{2}T_{(-\Delta+m^2)^{-1}j}(j)\right)$$
$$= \exp\left(-\frac{1}{2}\langle j, (-\Delta+m^2)^{-1}j\rangle_{L^2(R^D)}\right)$$
$$= \int d_{(-\Delta+m^2)^{-1}}\mu(\phi)(\exp i\phi(j)). \tag{7.3-c}$$

Interacting quntum field theories as perturbation around massive scalar fields, by theirs turn are also defined by an absolutely continuous measure $d\nu(\phi)$ in relation to previously euclidean massive free scalar field $d_{(-\Delta+m^2)^{-1}}\mu(\phi)$, which in a rigorous way would be writen as of as

$$d_{(\varepsilon)}\nu(\phi) = (d_{(-\Delta+m^2)^{-1}}\mu(\phi))(e^{-gV(\phi*\rho_\varepsilon)}) \tag{7.4}$$

where $(\phi * \rho_\varepsilon)$ denotes the field sampling in eq. (7.3-c) in its regularized form (which are $C^\infty(R^N)$ functions).

Classifically $gV(\phi)$ is a non-linear function of the C^∞-regularized distribution $(\phi * \rho_\varepsilon)$ and g is the bare theory's coupling constant.

Note that trying to formalize rigorously eq. (7.4) for $\varepsilon \to 0$ is meaningless, since one expect that if these probabilistic distributional field configurations were described by locally integrable functions, they would be infinite almost everywhere on R^D. It is no difficult to expect such behaviour for scalar neutral free euclidean fields since (for $D > 1$)[1]

$$\int d_{(-\Delta+m^2)^{-1}}\mu(\phi)\phi^2(x) = \lim_{x\to y}\int d_{(-\Delta+m^2)^{-1}}\mu(\phi)(\phi(x)\phi(y))$$
$$= \frac{1}{(2\pi)^{D/2}}\int d^Dk\frac{1}{k^2+m^2} = +\infty \tag{7.5}$$

As a consequence it does not make sense to consider interactiong euclidean QFT interactions directly from their classical counterpart Lagrangians as usually done in Physics textbooks. For instance, the functional below it holds true the triviality relation on the mentioned field measure space.

$$\exp\left\{-\frac{\lambda}{4!}\int \phi^4(x)d^Dx\right\} = 0 \quad \text{a.e. on} \quad (S'(R^D), d_{(-\Delta+m^2)^{-1}}\mu(\phi)). \tag{7.6}$$

[1]See Appendix B

For interacting field theories as defined as perturbation such "triviality" behavior is also expected.

In order to overcome such mathematical problems, it is used "regularized" forms of the QFT, which in many cases, although useful for computations, they shadow on the famous problems of infinities or the usual Feynman-Dhyson perturbative scheme (intrinsic for defining interaction field path measures around free field path integrals).

Let us now propose a new form of regularizing euclidean quantum field theories by considering suitable regularization on the Kinetic free operator eq. (3-b).

We have the following basic theorem of ours ([2]).

Theorem 1. *Let us consider the α-power of the Laplacean operator acting on $H^{2\alpha}(R^D)$. Let us consider also a compact domain $\Omega \subset R^D$ and $\chi_\Omega(\tau)$ its characteristic function*

$$\left(\chi_\Omega(x) = \begin{cases} 1 & x \in \Omega \\ 0 & x \notin \Omega \end{cases} \right)$$

then one has:

1 – the operator $\mathcal{L}_{(\alpha,\Omega,m)}$ here defined by its inverse which is given by an integral operator with kernel

$$\mathcal{L}^{-1}_{\alpha,\Omega,m}(x,y) = \chi_\Omega(x)[(-\Delta)^\alpha + m^2]^{-1}(x,y)\chi_\Omega(y) \tag{7.7}$$

is such that it defines a probability measure on the functional space $L^2(R^D)$ for the parameter range $\alpha > D/2$

$$\begin{aligned} Z(j) &= \exp\left\{ -\frac{1}{2}\langle j, \mathcal{L}^{-1}_{(\alpha,\Omega,m)} j\rangle_{L^2(R^D)} \right\} \\ &= \int_{L^2(R^D)} \left(d_{\mathcal{L}^{-1}(\alpha,\Omega,m)}\mu(\phi) \exp i\langle \phi, j\rangle \right). \end{aligned} \tag{7.8}$$

Proof: We note that $\mathcal{L}^{-1}_{\alpha,\Omega,m}$ is a positive definite trace class operator on $L^2(R^D)$ for $\alpha > \frac{D}{2}$ (ref. ([2])). Since

$$Tr_{L^2(R^D)}\{\mathcal{L}^{-1}_{(\alpha,\Omega,m)}\} = \frac{\mathrm{vol}(\Omega)}{(2\pi)^D}\left[\int_{R^D} \frac{d^D k}{k^{2\alpha} + m^2}\right] < \infty \quad \text{if} \quad \alpha > \frac{D}{2}. \tag{7.9}$$

Now the result on the support of the probability measure given by eq. (7.8) as given by the $L^2(R^D)$ (or $L^2(\Omega)$) space is a direct consequence of the Minlos-Bochner theorem on integration theory on Hilbert Spaces setting ([2] – appendix). □

The point now in consider finite volume analitically regularized free scalar QFT's is that one can handle directly non-trivial interactions (super renormalizable QFT's) since the field condigurations are now the usual real measurable point $L^2(\Omega)$ functions instead of L. Schwartz distributions.

For instance, we have the following theorem:

Theorem 2. *The exponential cut-off euclidean $P_2(\varphi)$ interaction (defined explicitly below) is well defined on the $(d_{\mathcal{L}^{-1}_{(\alpha,\Omega,m)}}\mu(\phi), L^2(\Omega))$ probability functional space for $\alpha > \frac{D}{2}$*

$$V_\Omega(\varphi) = \int_{\Omega \subset R^D} \exp(-\delta\varphi^2(x)) \left[\sum_{\substack{j=0 \\ j=even}}^{2k} \frac{\lambda_j}{j'} \varphi^j(x) \right] d^D x \tag{7.10}$$

here $\delta > 0$, $\lambda_j \geq 0$, for $j = 0, 2, 4, \ldots, 2k$.

Proof: An immediate consequence of the fact that if $\varphi(x) = +\infty$ for $\varphi \in L^2(\Omega)$ and for some point $x \in \Omega$, then $V(\varphi(x)) = 0$, which means that $e^{-V(\phi)} = 1$. As a result one has the upper bound for all

$$\phi \in L^2(\Omega) = \text{supp } d_{\mathcal{L}^{-1}_{(\alpha,\Omega,m)}}\mu(\varphi) \Rightarrow \exp\{-V_\Omega(\phi)\} \leq 1. \tag{7.11}$$

Since $(d_{\mathcal{L}^{-1}_{(\alpha,\Omega,m)}}\mu(\phi))$ is a truly functional probability measure for $\alpha > \frac{D}{2}$, an application of the Lebesgue dominated theorem gives the finitude of the associated generating functional eq. (7.8) for functional interactions of the form eq. (7.10). □

Note that remains a non trivial problem to evaluate the n-point field correlation functions on this proposed scheme of ours. However the same reasoning below can be applied to prove that the exponential regularized n-point functions are finite. Namelly

$$\begin{aligned} &\int_{L^2(\Omega)} (d_{\mathcal{L}^{-1}_{\alpha,\Omega,m}}\mu(\phi)) \exp\{-V_\Omega(\phi)\} \\ &\quad \times \exp\left(i \int_\Omega (e^{-\delta\phi^2}\phi)(x) j(x) d^D x \right) \\ &\quad = Z_{(\alpha,\Omega,m^2,\delta)}(j(x)) \end{aligned} \tag{7.12}$$

is an analytical functional on $L^2(\Omega)$. □

Theorem 3. *The analytic regularized, finite volume generating functional* $Z_{\alpha,\Omega,m}[j]$ *is defined now on the Sobolev Space* $H_0^m(\Omega)$ *(now* Ω *denoting an open set with compact closure) if* α *is sufficiently higher* $(2\alpha - 2m - D > 0)$ *and for Drichlet conditions on the field configurations on the* Ω*-boundary* $(\phi|_{\partial\Omega} \equiv 0)$.

On the light of such theorem, one can use the Sobolev immersion Theorem $H_0^m(\Omega) \subset C^P(\Omega)$, for $P < m - D/2$, to have continuous or even differentiable euclidean field sample configurations on the theory's path integral for higher order free field path integral.

At this point appears the very difficult problem of cut-off remotions $\text{vol}(\Omega) \to \infty$, $\alpha \to 1$, $\delta^2 \to 0$ or and $m^2 \to 0$ in these analitically "regularized" field thories. In next section one implement the finite volume and the analitically regularized remotion on a class of non trivial massless scalar field theories, just for exemplifying that cutt-off remotions on ours proposed regularized path integrals are as possible as well.

Let us point out that theorem 3 can be considered as a sort of generalized Wiener theorem on the continuity of Brownion motions for ours volume-analitically regularized euclidean fields path integrals.

The proof of Theorem 3 is again a direct result of the Minlos-Bochner theorem ([1], [2], [6]).

$$\int_{L^2(R^D)} d_{\mathcal{L}^{-1}_{\alpha,\Omega,m}} \mu(\phi) ||\varphi^2||_{H^m(\Omega)} < \infty \tag{7.13-a}$$

if

$$Tr_{H^m(\Omega)}[\mathcal{L}^{-1}_{\alpha,\Omega,m}] < \infty. \tag{7.13-b}$$

This can be verified by a direct computation

$$Tr_{H^m(\Omega)}[\mathcal{L}^{-1}_{\alpha,\Omega,m}] = \frac{1}{\text{vol}(\Omega)} \left\{ \int_{R^D} d^D k \int_{R^D} d^D u \frac{k^{2m}}{u^{2\alpha} + m^2} |\hat{I}_\Omega(k-U)|^2 \right\} < \infty \tag{7.13-c}$$

if $m + D < \alpha$.

The reader can check eq. (7.13-c) by means of the finitude condition of the integral [2]

$$\int d^D k \frac{K^{2m}}{K^{2\alpha} + m^2} < \infty \quad \text{if} \quad 2\alpha - 2m - 2D > 0. \tag{7.13-d}$$

[2]See also Appendix B for a discussion on the sample differentiability of the Euclidean Field Theory Path Integraly on Hilbert Spaces.

7.3 On the cut-off remotion on a two-dimensional Euclidean QFT model

We start our studies in this section by considering the (bare) euclidean functional integral on a finite volume smooth compact region $\Omega \subset R^2$ as given below (see ref. [2]) for $\alpha > 1$, associated to a real scalar field on $\mathbb{R}^2$

$$
\begin{aligned}
Z_\alpha(j(x)) = \frac{1}{Z_\alpha(0)} &\left\{ \int_{L^2(\Omega)} (d_{\mathcal{L}_\alpha^{-1}}\mu)(\varphi) \right\} \\
&\times \exp\left(-g_{\text{bare}} \int_{\mathbb{R}^2} V(\varphi(x)) d^2x \right) \\
&\times \exp\left(i \int_{\mathbb{R}^2} j(x)\varphi(x) d^2x \right). \qquad (7.14)
\end{aligned}
$$

Here the functional measure on the path space of real square integrable function on Ω, denoted by $L^2(\Omega)$ is given through the Minlos Theorem (see Section 1) for real field sources $j(x)$ on $L^2(R^2)$ and the parameter α on the range $\alpha > 1$. Here m^2 is a mass parameter eventually vanishing at the end of our estimate, since we are only interested on the ultraviolet field singularities

$$
\begin{aligned}
&\exp\left\{ -\frac{1}{2} \int_{\mathbb{R}^2 \times \mathbb{R}^2} d^2x d^2y \; j(x)(\chi_\Omega(x)((-\Delta)^{-\alpha} + m^2)(x,y)\chi_\Omega(y))j(y) \right\} \\
&= \int_{L^2(\Omega)} (d_{\mathcal{L}_\alpha^{-1}}\mu)(\varphi) \exp(i\langle j, \varphi \rangle_{L^2(R^2)}). \qquad (7.15)
\end{aligned}
$$

The interaction potential is a continuous function vanishing at infinite such that it posseses an essential bounded $L^1(R)$ Fourier Transform (for instance $V(x) = \frac{\lambda}{4!} e^{-\delta x^2} x^4$, etc...). In others words:

$$
||V||_{L^\infty(\Omega)} \leq ||\widetilde{V}||_{L^\infty(R)} < \infty. \qquad (7.16)
$$

Let us show that by defining the bare coupling constant by the renormalization prescriptions ($v = \text{vol}(\Omega)$)

$$
g_{\text{bare}}(\alpha, v) = \frac{g_{\text{rem}}}{(1-\alpha)^{1/2} v} \qquad (7.17)
$$

The functional integral eq. (7.1) has a finite limite at $\alpha \to 1$, $v \to \infty$, when understood in the R.P. Feynman sense as an expansion perturbative on the renormalized constant g_{rem}. Namely (see Chapter 5, §5.2, eq. (5.11), [7])

$$
Z_{\alpha=1}(j(x)) = \lim_{V\to\infty} \left(\lim_{N\to\infty} \left(\lim_{\alpha\to 1} \left(\lim_{m^2\to 0} I_N(g_{\text{bare}}(\alpha, v), [j]) \right) \right) \right). \qquad (7.18)
$$

Theorem 1. *The functional $I_N(g_{bare}(\alpha, v), [j])$ satisfies the upper bound at the limit $\alpha \to 1$*

$$\lim_{N\to\infty}\left(\lim_{\alpha\to 1}|I_{N,\alpha}(g_{bare}(\alpha, v), [j]|)\right) \leq C^2\exp(C) \tag{7.19-a}$$

where the constant C is given by

$$C = (4\pi)^{1/2} g_{ren}||\widetilde{V}||_{L^\infty(R)}. \tag{7.19-b}$$

Proof: By noting that $\int_{C(\Omega)}(d_{\mathcal{L}_\alpha^{-1}}\mu)(\varphi)\exp(ik\varphi(x)) = 0$, one has the following result eq. (7.8) (see [2]), where the integral kernel of the our "free" propagator is given explicitly by (for $\alpha > 1$; see Appendix A)

$$\mathcal{L}^{-1}_{\alpha,m^2=0}(x_i,x_j) = \chi_\Omega(x)\left[\left(\frac{1}{(2\pi)}(|x_i-x_j|^{2(\alpha-1)})\left(\frac{\Gamma(1-\alpha)}{\Gamma(\alpha)}2^{2(1-\alpha)}\right)\right)\right]\chi_\Omega(y).$$

$$\begin{aligned} I_{N,\alpha}&(g_{\text{bare}}(\alpha, v), [j]) \\ &= \sum_{n=0}^{N}\left\{\frac{(-1)^n}{n!}(g_{\text{bare}}(\alpha, v))^n\int_\Omega d^2x_1\ldots d^2x_n\right. \\ &\times\int_R\frac{dk_1}{(2\pi)^{1/2}}\cdots\int_R\frac{dk_n}{(2\pi)^{1/2}}(\widetilde{V}(k_1)\cdots\widetilde{V}(k_n)) \\ &\times\left[\int_{C(\Omega)}(d_{\mathcal{L}_\alpha^{-1}}\mu)(\varphi)\exp\left(\sum_{\ell=1}^{n}ik_\ell\varphi(x_\ell)\right)\right. \\ &\times\left.\left.\exp\left(i\int_\Omega j(x)\varphi(x)d^2x\right)\right]\right\} \end{aligned} \tag{7.20}$$

As a consequence of the positiviness of the "kinetic" Green function $\mathcal{L}_\alpha^{-1}(x,y)$; one has

$$\begin{aligned} (I_{N,\alpha}&(g_{\text{bare}}(\alpha, v), [j]) \leq 1 \\ &+\left\{\sum_{n=1}^{N}\frac{(|(g_{\text{bare}}(\alpha, v))|)^n}{n!}(||\widetilde{V}||_{L^\infty(\Omega)})^n\int_\Omega d^2x_1\ldots d^2x_n\right. \\ &\times\left.\left[\det_{N\times N}^{-1/2}[\mathcal{L}^{-1}_{\alpha,m^2}(x_i,x_j)]\right]\right\}. \end{aligned} \tag{7.21}$$

Note that due to the continuity on the infrared cut-off mass parameter, it is possible to consider directly its limit on the determinant formed by the Green's functions.

At this point we note the Taylor expansion of the below written object

$$L_N(\alpha,v) = \int_\Omega d^2x_1 \cdots d^2x_n \underset{\substack{N\times N 1\le i\le N\\ 1\le j\le n}}{\det}{}^{-1/2} [\mathcal{L}_\alpha^{-1}(x_i,x_j)]$$
$$= (1-\alpha)^{N/2} C_n + (1-\alpha)^{\frac{N}{2}+m_1} C_{N+1} + \dots \quad (7.22)$$

with

$$C_n = v^n (4\pi)^{N/2} \left(\underset{\substack{1\le i\le N\\ 1\le j\le N}}{\det} [A_{ij}] \right)^{-1/2} \quad (7.23\text{-a})$$

and the matrix $[A_{i,j}]$ is defined by the rule

$$[A[_{ij} = \begin{cases} 0 & \text{if} \quad i = j \\ 1 & \text{if} \quad i \neq j \end{cases} \quad (7.23\text{-b})$$

It yield thus ([2], [3]) for $N > 1$

$$\lim_{V\to\infty} \left(\lim_{\alpha\to 1} |(g_{\text{bare}}(\alpha,v) L_N(\alpha,v))| \right.$$
$$= \lim_{V\to\infty} \left\{ \lim_{\alpha\to 1} \left[\left| \left(\frac{g_{\text{ren}}}{(1-\alpha)v} \right)^N L_N(\alpha,v) \right| \right] \right.$$
$$= \lim_{V\to\infty} \left(\lim_{\alpha\to 1} \left\{ \left(\frac{|(g_{\text{ren}})|^N}{v^N |(1-\alpha)|^{N/2}} \right) \right. \right.$$
$$\times \left[|(1-\alpha)|^{N/2} V^N (4\pi)^{N/2} (|(-1)(N-1)(-1)^N|)^{-1/2} \right] \Big\} \Big)$$
$$= (|g_{\text{ren}}|)^N \frac{1}{(N-1)^{1/2}} ((4\pi)^{1/2})^N \Big) \quad (7.24)$$

We have thus the uniform bound on the "interaction order" N in our Euclidean QFT model

$$\lim_{V\to\infty} \lim_{\alpha\to 1} (|I_{N,\alpha}(g_{\text{bare}}(\alpha,v),[j])|)$$
$$\le 1 + \left\{ \sum_{N=2}^{M} \frac{(4\pi)^{N/2}}{N!} \frac{(|g_{\text{ren}}|)^N}{(N-1)^{1/2}} \right.$$
$$\times \left. [\||\widetilde{V}\|_{L^\infty(\Omega)}]^N \right\}$$
$$\le C^2 \left(\sum_{N=0}^{\infty} \frac{C^N}{N!} \right)$$
$$= C^2 \exp(C) \quad (7.25)$$

with

$$C = |g_{\text{ren}}| \cdot ||\widetilde{V}||_{L^\infty}(4\pi)^{1/2}. \tag{7.26}$$

Note that we have used the elementary estimate to arrive at eq. (7.25) for $N > 1$

$$\frac{1}{(N-1)^{1/2}} \leq 1, \quad \text{for } N \geq 2. \tag{7.27}$$

We conclude this, that the functional path integral eq. (7.14) under the renormalization coupling constant eq. (7.17) and rigorous Feynman perturbative definition eq. (7.18) has a finite limit for $\alpha = 1$.

It is worth that one could also consider the most general multiplicative renormalization including the functional form of the interaction

$$g_{\text{bare}}(\alpha, v||\widetilde{V}||_{L^\infty(\Omega)}) = \frac{g_{\text{ren}}}{(1-\alpha)^{1/2} v||\widetilde{V}||_{L^\infty(\Omega)}}. \tag{7.28}$$

Now allowing interactions satisfying the constraint $\widetilde{V}(k) = \lim_{\ell\to\infty} \widetilde{V}_\ell(k)$ with $||\widetilde{V}_\ell(k)||_{L^\infty(\Omega)} = \ell \in \mathbb{N}^+$.

It is worth to recall that we have proven that the full generating functional eq. (7.14) at $\alpha \to 1$ as defined by a Feynman's perturbative series: Feynman's diagrammas renormalized order by order in a power serie expansion on the bare coupling constant is finite and it is a continuous functional on the source space $j(x) \in L^2(\Omega)$.

However it appear that the use of the propagator prescription

$$\begin{aligned} \hat{\mathcal{L}}^{-1}_{\alpha,m^2=0}(x_i, x_j) &= \chi_\Omega(x)\Big(\Big[\frac{1}{4\pi}\frac{\Gamma(1-\alpha)}{\Gamma(\alpha)}2^{2(1-\alpha)} \\ &\quad \times |x-y|^{2(\alpha-1)}\Big] - \frac{4\pi}{(1-\alpha)}\Big)\chi_\Omega(y), \end{aligned} \tag{7.29}$$

which converges on the $D'(R^2)$ L. Schwartz distributional sense to the usual non positive definite two-dimensional Laplacean Green function for $\alpha \to 1$ does not lead to well defined Euclidean QFT generating functional. A result already expected since Massless $2D$ Euclidean Q.F.T. Theories built already as perturbation around free scalar Massless fields on R^2 do not make mathematical sense due to the fact that the two-dimensional Laplacean Green function does not belongs to the "Fourier Transformable" Tempered Distributional Space $S'(R^2)$, a fact already observed a long time ago by S. Coleman ([4]) and fully used by G. Hoft on his studies on $(QCD)_2$ – solubility at large number colors ([5]).

Another point worth call attention in this Section is that the same proof works out for a class of four-dimensional analitically regularized Euclidean

Field theories with the "Free kinetic operator" defined through the Minlos's theorem on a finite volume region $\Omega \subset R^4$

$$\exp\Big\{-\frac{1}{2}\int_{R^4} d^4x \int_{R^4} d^4y \\ \times j(x)\Big\{\chi_\Omega(x)\Big[(-\Delta^2)^{-\alpha}+m_0^2\Big]\chi_\Omega(y)\Big\}j(y) \\ = \int d_{\mathcal{L}^{-1}_{\alpha,m_0^2}}\mu(\varphi)\exp\Big(i\int_{R^4}\varphi(x)j(x)d^4x\Big)\Big\}. \quad (7.30\text{-a})$$

Here the Integral Kernel of the square D-fimensional Laplacean is given by (for $\alpha > 1$)

$$(-\Delta^2)^{-\alpha} = \frac{\Gamma(\frac{D}{2}-2\alpha)}{\Gamma(2\alpha)2^{4\alpha}\pi^{D/2}}(|x-y|^{4\alpha-D}). \quad (7.30\text{-b})$$

Finally, we call attention that into another publication we will address to the "differentiability" of the generating functional eq. (7.1) at $\alpha \to 1$ as defined in the Bulk of this section. However it is straithforward to obtain such differentiability for sources $j(x)$ coupled to field configurations interaction of the form $\exp(-\delta\varphi^2(x))\varphi(x)$. Note that in this case, the N-point Taylor's coeficients of $Z[j(x)]$ are explicitly given by

$$\frac{\delta^N Z[j(x)]}{\delta j(x_1)\cdots\delta j(x_N)}\Big|_{j(x)\equiv 0} \\ = \int_{C(\Omega)}(d_{\mathcal{L}_\alpha^{-1}}\mu)(\varphi)\left(\prod_{\ell=1}^{N}[\exp(-\delta\varphi^2(x_\ell))\varphi(x_\ell)]\right) \\ < \infty, \quad (7.31)$$

since the domain of the above functional integral for $\alpha > 1$ is the space of measurable square integrable functions on Ω (and for $\delta > 0$)

$$||e^{-\delta\varphi^2(x)}\varphi(x)||_{L^\infty(R)} = (\max_{x\in R}|e^{-\delta x^2}x|) = C < \infty, \quad (7.32)$$

leading to the finitiness of eq. (7.31) by the use of the Lebesgue dominated convergence theorem.

The limite of $\delta \to 0$ on the momentums eq. (7.31) will appears elsewhere.

Finally we wishe to point out that non trivial homological topology of the compact planar two-dimensional domain Ω ([8]) in ours path integral can be easily taken into account by the Ω set indicator function $\chi_\Omega(\Omega)$ on eq. (7.30-a) of this section, specially on Fourier Space by means of the Ω-domain Fourier Integral form factor for Ω with holes inside

$$\hat{I}_k(\Omega) = \left(\int_\Omega d^2\xi \exp(ik\xi)\right); \quad (7.33)$$

which appears on the expression of the theory's propagator on momentum space for general R^D space-time

$$\hat{\mathcal{L}}^{-1}(k,k') = \int_{R^D} \left(\frac{\hat{I}_\Omega(k-p)\hat{I}_\Omega(p-k')}{p^{2\alpha}+m^2} \right) d^D p \tag{7.34}$$

and leading thus to the Feynman diagrammotic generating functional on the Fourier Space

$$\begin{aligned} &Z[\tilde{j}(k)]/Z[0] \\ &\quad = \exp\left\{ -\int_\Omega d^D x \left(V\left(\frac{1}{(2\pi)^{D/2}} \int_{R^D} d^D k e^{+ikx} \left(\frac{\delta}{\delta \tilde{j}(k)} \right) \right) \right) \right\} \\ &\quad \times \exp\left\{ -\frac{1}{2} \int_{R^{2D}} dk dk' \tilde{j}(k) \right. \\ &\quad \times \left. \left(\int_{R^D} dp \frac{\hat{I}(k-p)\hat{I}(p-k')}{p^{2\alpha}+m^2} \right) \tilde{j}(k') \right\} \end{aligned} \tag{7.35}$$

As a last remark, we conjecture that the ultra-violet limit $\alpha \to 1$ on the usual correlations functions associated to our path integral should expected to be finite. The argument follows by considering $||\widetilde{V}||_{L^\infty(\Omega)} = 1$, since $x \to 0$ and thus, obtaining the general structure of the (for instance) two-point function at perturbative order N

$$\begin{aligned} \langle \varphi(x_1)\varphi(x_2) \rangle \overset{\alpha\to 1}{\sim} &-(\mathcal{L}^{-1}_{\alpha,m^2=0}(x_1,x_2)) \\ &+ \sum_{p=1,q=1}^{N} \left\{ \mathcal{L}^{-1}_{\alpha,m^2=0}(x_1,x_p) \right. \\ &\quad \times \left. [\mathcal{L}^{-1}_\alpha(x_i,x_j)]^{-1}_{pq} \mathcal{L}^{-1}_{\alpha,m^2=0}(x_q,x_2) \right\} \end{aligned} \tag{7.36}$$

and noting the Laplace formula for evaluate the inverse of the propagator matrix

$$\begin{aligned} [\mathcal{L}^{-1}_\alpha(x_i,x_j)]^{-1}_{pq} &= \frac{1}{\det_{N\times N}[\mathcal{L}^{-1}_\alpha(x_i,x_j)]} \\ &\quad \times (\mathbb{C}(x_i,x_j)]_{qp}, \end{aligned} \tag{7.37}$$

with the cofactor matrix $[\mathbb{C}(x_i,x_j)]_{qp}$ associated to the $[\mathcal{L}^{-1}_\alpha(x_i,x_j)]$ propagator matrix eq. (7.21). One expects thus that the singular behavior for $\alpha \to 1$ of the determinant

$$\det_{N\times N}[\mathcal{L}^{-1}_\alpha(x_i,x_j)] \overset{\alpha\to 1}{\sim} (1-\alpha)^{-N},$$

cancels out with the factor

$$\sum_{p=1,\varepsilon=1}^{N} \mathcal{L}^{-1}_{\alpha,m^2}(x_1,x_p)[\mathbb{C}(x_i,x_j)]_{qp}\mathcal{L}^{-1}_{\alpha,m^2=0}(x_q,x_2) \sim (1-\alpha)^{-N},$$

on eq. (7.36).

7.4 On the construction of the Wiener Measure

On next Section 4, we intend to analyze the somewhat different functional-path integral on functional space, mainly due R.P. Feynmann and M. Kac: the so called geometrodynamical end points fixed field propagator.

However, such objects to be defined mathematically, one must review the construction of the famous Wiener path measure ([6]–[7]). This is our objective in this short section.

Let us first introduce some notations and mathematical objects.

We first write the time fixed Heat equation green function on the one-point compactified of the real line R, the interval $[-\frac{\pi}{2}, \frac{\pi}{2}]$ as a integral kernel of a continuous linear functional on the compact support continuous function f on R. So, let $\varepsilon > 0$ and $f \in C_c(R)$

$$L_\varepsilon(f) = \int_R \overbrace{\left(\frac{\exp(-|x-y|^2/2\varepsilon)}{(2\pi\varepsilon)^{1/2}}\right)}^{:=\hat{G}_0(x,y,\varepsilon)} f(y)dy$$

$$= \int_{-\frac{\pi}{2}}^{\frac{\pi}{2}} \left(\frac{e^{-|tg(\theta_x)-tg(\theta_y)|^2/2\varepsilon}}{(2\pi\varepsilon)^{1/2}\cos^2\theta}\right) f(tg\theta)d\theta. \tag{7.38}$$

Note that supp $f(tg\theta) \subset (-\frac{\pi}{2}, -\frac{\pi}{2})$. For a given $g(\theta) \in C_c([-\frac{\pi}{2}, \frac{\pi}{2}])$, eq. (7.38) defines a positive continuous linear functional on $C_0(\dot{R})(f(\infty) = g(-\frac{\pi}{2}) = g(\frac{\pi}{2}) = 0)$.

Let us define the following projective family a positive linear functionals on $C_c(\prod_{n=0}^{\infty}([-\frac{\pi}{2}, \frac{\pi}{2}])_n)$, firstly defined on the σ-algebra of the infinite variable space $C_c(\prod_{n<0}^{\infty}\dot{R}) \equiv C_0((\dot{R})^\infty)$. For N a given positive integer fixed, but arbitrary and $\bar{x}$ a fixed point on $[-\frac{\pi}{2}, \frac{\pi}{2}]$, we consider the projected positive continuous linear functionals $(\varepsilon = \frac{1}{N})$;

$$L_{\bar{x}}^{(N)}(f(x_1,\dots,x_N)) = \int_{\dot{R}} dx_1 \dots \int_{\dot{R}} dx_N (f(x_1,\dots,x_N))\Big) \\ \times (\hat{G}_0(x,x_1,\varepsilon)\dots\hat{G}_0(x_{N-1},x_N,\varepsilon))dx_1\dots dx_N. \tag{7.39}$$

We point out the "projective" properties of the family of positive continuous functionals $\{L_{\bar{x}}^{(N)}\}$:

a) For $M \le N$,

$$L_x^{(M)}(f(x_1,\dots x_M)) = L_x^{(N)}(f(x_1,\dots,x_M)) \tag{7.40}$$

b) $C_{c,\text{finite variables}}(\dot{R}^\infty) = \{f \in C_c(\dot{R}^\infty)$, but with finite variables$\}$ is a dense subset of $C_c(\dot{R}^\infty)$ (endowed with the usual supremum norm!).

All theses results above remarked show that exists the

$$\lim_{N\to\infty} L_{\bar{x}}^{(N)} = L_{\bar{x}}^{(\infty)} \quad \text{on } C_c(\dot{R}^\infty).$$

Since $\dot{R}^\infty$ is a compact topological space one can apply the Riesz Markov theorem to represent $L_{\bar{x}}^\infty$ through a well defined measure on $\dot{R}^\infty$ (the Bare σ-albegra of $\dot{R}^\infty$).

Note that one could take $\varepsilon = t/N$ with $t > 0$, a real fixed, and this obtain the famous Wiener measure ending at $\bar{x}$ at time t

$$L_{(\bar{x},t)}^\infty(f) = \int_{\dot{R}^\infty} d_{(\bar{x},t)}^{\text{Wiener}} \hat{\mu}[g(\sigma)] f(g(\sigma)) \tag{7.41}$$

where $g(\sigma) \in \dot{R}^\infty$ is identified with the set of all real functions on $\dot{R}$ [the "compactified" Wiener path trajectory], with the domain $\sigma \in [0, t]$.

It is worth to remark that on eq. (7.39), all the "time parameters" are at the same value $t = \varepsilon$.

It is an open problem to show the existence and unicity of the Wiener measure $d_{(\bar{x},t)}\mu[g(\sigma)]$ under general (different) time steps on eq. (7.39).

It is worth also to note that $f \in C_c(\dot{R}^\infty, R)$ by the hypothesis of the Riesz-Markos theorem ([7]).

At this point if is argued that there is a unique "pull-back" of the above constructed Wiener measure on the space of compact paths to the full R paths. Namely, for $F \in C_c(C(R,R),R)$ and $x \in R$

$$L_{(x,t)}^\infty(F) = \int d_{(x,t)}^{\text{Wiener}} \mu(X(\sigma)) F(X(\sigma)). \tag{7.42}$$

The above construction generalizes straightforwardly for R^D $(D > 1)$.

We have thus the following theorem (Feynmann Wiener-Kac): Let $(-\Delta)$ be the essential self-adjoint extension of the usual Laplacean acting on $C_c^\infty(R^D)$.

We have the formula (7.7), for $F \in C_c(R^D)$ on the sense the topology of $C_c(R^D)$

$$\begin{aligned}(e^{-\frac{t}{2}\Delta} F)(x) \\ = \int (d_{(x,t)}^{\text{Wiener}} \mu(X(\sigma)) F(X(t)).\end{aligned} \tag{7.43}$$

For general $F \in L^2(R^D)$, eq. (7.43) is obtained by (unique) extension, since $((\overline{C_c(R^D)})_{L^2(R^D)} = L^2(R^D))$.

It is important to call attention that due to the C^∞-regularizing property of the heat Kernel eq. (7.38), the functional integral representation eq. (7.43) remains correct in the $L^2(R^D)$ sense for $F(w) = \delta^{(D)}(w-y)$ and leading thus to the formal Brownian Bridge path integral measure representation for the Heat Kernel

$$\langle x|e^{-\frac{t}{2}\Delta}F|y\rangle \overset{L^2(R^D)}{=} \int_{X(0)=y}^{X(t)=y} d^{\text{Wiener}}\mu(X(\sigma))F(X(t)) \qquad (7.44\text{-a})$$

or in the correct mathematical meaning of the above written eq. (7.44-a) for f and $g \in L^2(R^D)$

$$\int_{R^D} f(x)\langle x|e^{-\frac{t}{2}\Delta}F|y\rangle\bar{g}(y) = \int_{X(0)=y}^{X(t)=y} d^{\text{Wiener}}\mu(X(\sigma))(f(X(t))\bar{g}(X(0))). \qquad (7.44\text{-b})$$

7.5 On the Geometrodynamical Path Integral

Sometimes it appears to be useful for calculational purposes on euclidean quantum field theory to give a generalized meaning for the Brownian Bridge Wiener path integral eqs. (7.44-a)–(7.44-b), called now the Geometridynamical propagator connecting a classically observed field configuration $\phi(x,t_1) = \beta_{in}(x)$ to another final one $\phi(x,t_2) = \beta_{out}(x)$, with $t_2 > t_1$.

Let us formulate the problem for the free case of a real scalar field $\phi(x,t)$ with classical action and with Dirichlet boundary conditions on the D-dimensional space time cylinder manifold propagation $D = \Omega \times [t_1,t_2]$ with A denoting an inversible positive definite self-adjoint elliptic operator on Ω

$$S[D] = \int_D \frac{1}{2}\left(\beta\left(-\frac{\partial^2}{\partial t^2} + A\right)\beta\right)(x,t)d^{D-1}xdt. \qquad (7.45)$$

One wants to give a rigorous mathematical meaning for the Euclidean Feynman-Wheller path integral

$$G[(\beta_{in}(x),t_1);(\beta_{out}(x),t_2)] = \int_{\beta(y,t_1)=\beta_{in}(x)}^{\beta(x,t_2)=\beta_{out}(x)} D^F[\beta(x,t)]\exp\{-S[D]\}. \qquad (7.46)$$

The most usual way to give a mathematical meaning for eq. (7.46) is to use the spectral theorem for A ($A\varphi_\mu = \lambda_\mu \varphi_\mu$) and regard eq. (7.46) as the (enumerable) infinite product of Brownian Bridge Wiener measures eq. (7.44-a) and under the hypothesis that all the field configurations entering on the support of the resulting field path integral measure is of the form

$$\beta(x,t) = \sum_{\mu=0}^{\infty} C_\mu(t)\phi_\mu(s) \in C([t_1,t_2], L^2(\Omega)). \tag{7.47}$$

We thus define eq. (7.46) as

$$G[(\beta_{in}(x),t_1),(\beta_{out}(x),t_2)]$$

$$= \prod_{n=0}^{\infty}\left[\int_{X_n(t_1)=\beta_n^{in}}^{X_n(t_2)=\beta_n^{out}} d^{\text{Wiener}}\mu(X_n(\sigma)) \times \exp\left(-\frac{1}{2}\int_{t_1}^{t_2}(\lambda_n)^2(X_n(\sigma))^2\right)\right]. \tag{7.48}$$

Here

$$\beta^{in}(x) = \sum_{n=0}^{\infty} \beta_n^{in}\varphi_n(x) \tag{7.49-a}$$

$$\beta^{out}(x) = \sum_{n=0}^{\infty} \beta_n^{out}\varphi_n(x). \tag{7.49-b}$$

Let us note that the enumerable infinite product of Wiener-Harmonic Oscilator measure is still a well behaved σ-measure on the product measure space $\prod_{n=0}^{\infty}(C([t_1,t_2],R))_n$. Note that if one user the compactified of the real line as in Section 3, one would gets as the measure space; the compact space $\prod_{n=0}^{\infty} C([t_1,t_2],\dot{R})_n$.

In the presence of an external source $f(t,x) \in C([0,T], L^2(\Omega))$, one has the usual Feynman closed expression in terms of Feynman-Wiener notation for the Wiener-Harmonic oscillator path measures ([7])

$$G[(\beta^{in}(x),0);(\beta^{out}(x),T),[j(x,t)]]$$

$$= \prod_{n=0}^{\infty} \int_{X_n(0)=\beta_n^{in}}^{X_n(T)=\beta_n^{out}} \left\{ D^F[X_n(\sigma)]\exp\left[-\frac{1}{2}\int_0^T \left(X_n\left(-\frac{d^2}{d\sigma^2} + \lambda_n^2 \right) X_k \right)(\sigma) \right] \right\}$$

$$\times \exp\left(\int_0^T d\sigma j_n(\sigma) X_n(\sigma) \right)$$

$$= \prod_{n=0}^{\infty} \left\{ \sqrt{\frac{\lambda_n}{Sinh(\lambda_n T)}} \right.$$

$$\times \exp\left\{ -\frac{\lambda_n}{2sinh(\lambda_n T)} \left[(\beta_n^{out})^2 \right. \right.$$

$$\left. \left. + (\beta_n^{in})^2 cosh(\lambda_n T) - 2\beta_n^{out}\beta_n^{in} \right] \right\}$$

$$- \frac{2\beta_n^{out}}{\lambda_n} \int_0^T d\sigma j_n(\sigma) sinh(\lambda_n \sigma)$$

$$- \frac{2\beta_n^{in}}{\lambda_n} \int_0^T d\sigma j_n(\sigma) sinh(\lambda_n (T-\sigma))$$

$$- \frac{2}{(\lambda_n)^2} \int_0^T d\sigma \int_0^T d\sigma' j_n(\sigma) j_n(\sigma') sinh(\lambda_n (T-\sigma))$$

$$\left. \times\, sinh(\lambda_n(\sigma')) \right\} \tag{7.50}$$

Another more attractive prescription to eq. (7.46), specially useful on String Theory ([7]) is to suppose that the sample space for geometrodynamical propagation is composed of field configurations made by random perturbations of a (fixed) classical field configuration as exposed in §5.3, eqs. (7.41)–(7.47) of ref. [7]. However, this method does not appears to be canonically invariant, since all the objects on the theory are dependent of the choosen background field configuration the classical choose field configuration, i.e. for different background field configuration one could obtain different path integrals.

7.6 Appendix A

Let us recall the following integral form of a Fourier Transform of a Tempered distribution T_f defined by a $L^2_{loc}(R^D)$ radial function $f(r)$

$$\mathcal{F}(T_{f(r)}) = T_{\hat{F}(k)} \tag{A-1}$$

with

$$\hat{F}(k) = (2\pi)^{D/2}\left(\int_0^\infty \frac{f(r)r^{D-1}J_{\frac{D-2}{2}}(kr)dr}{(kr)^{\frac{N}{2}-1}}\right). \tag{A-2}$$

By using our proposed distributional sense integral formulae for μ and ν complex numbers and $a > 0$

$$\int_0^\infty x^\mu J_\nu(ax)dx = 2^\mu a^{-\mu-1}\frac{\Gamma(\frac{1}{2}+\frac{1}{2}\nu+\frac{1}{2}\mu)}{\Gamma(\frac{1}{2}+\frac{1}{2}\nu-\frac{1}{2}\mu)}. \tag{A-3}$$

One obtains the result on the $S'(R^D)$ sense for $\alpha \in \mathbb{C}$

$$\mathcal{F}\left[T_{(\frac{1}{2\pi}\frac{\Gamma(1-\alpha)}{\Gamma(\alpha)}2^{2(j-\alpha)}r^{2(\alpha-1)}}\right]$$
$$= T_{K^{-2\alpha}}. \tag{A-4}$$

The complete distributional sense is given below for $f(x) \in S(R^D)$, with $\hat{f}(k) = \mathcal{F}[f(x)]$

$$T_{(\frac{1}{2\pi}\frac{\Gamma(1-\alpha)}{\Gamma(\alpha)}2^{2(1-\alpha)}r^{2(\alpha-1)}}(f) = T_{k^{-2\alpha}}(\hat{f}(k)). \tag{A-5}$$

Just for completeness, let us evaluate on the $S'(R^2)$ sense the Fourier Transform below

$$\begin{aligned} G_\alpha(x,y,m^2) &= \left(\frac{1}{\sqrt{2\pi}}\right)^2\left(\int_{R^2} d^2k e^{ik(x-y)}\frac{1}{(k^2+m^2)^\alpha}\right) \\ &= \frac{1}{2\pi}\left(\int_0^\infty dk\frac{k}{(k^2+m^2)^\alpha}\overbrace{(J_0(kr)+J_0(-kr))}^{2J_0(kr)}\right) \\ &= \frac{1}{2\pi}r^{2(\alpha-1)}\times\left(\int_0^\infty dp\frac{pJ_0(p)}{(p^2+m^2r^2)^\alpha}\right) \\ &= \frac{(mr)^{1-\alpha}\cdot K_{1-\alpha}(mr)}{2^{\alpha-1}\Gamma(\alpha)} \end{aligned} \tag{A-6}$$

where we have used the distributional sense integral relation for μ and ν complex parameters and $a, b \geq 0$:

$$\int_0^\infty \frac{J_V(bx)x^{\nu+1}}{(x^2+a^2)^{\mu+1}}dx = \frac{a^{\nu-\mu}b^\mu K_{\nu-\mu}(ab)}{2^\mu\mu(\mu+1)}. \tag{A-7}$$

Just for completeness, one can use the above exposed formulae to obtain the Integral Kernel of the $S'(R^D)$ distribution $(-\Delta)^{-\alpha}$. Namely

$$(-\Delta)^{-\alpha}(x,y) = \frac{\Gamma(\frac{D}{2}-\alpha)}{\Gamma(\alpha)2^{2\alpha}\pi^{D/2}(x-y)^{D-2\alpha}}. \tag{A-8}$$

7.7 Appendix B

Theorem. *Let A be a positive definite trace class operator on $L^2(\Omega)$, with spectral resolution $A\varphi_n = \lambda_n \varphi_n$ such that $\sum_{n=0}^{\infty} \lambda_n n^{2p} = +\infty$ for $p > 0$. Let also $d_A\mu(\varphi)$ denotes the cylindrical measure asociated to A through the Minlos-Bochner theorem applied to the bilinear source function $Z(j) = \exp\{-\frac{1}{2}\langle j, A_j\rangle_{L^2(\Omega)}\}$.*

Let $H_p = \{f \in L^2(\Omega) \mid f = \sum c_n \varphi_n$ with $\sum_{n=0}^{\infty} c_n^2 n^{2p} < \infty\}$ be the "Sobolev" sequence measurable sub-sets of $L^2(\Omega)$. Then we have that for any $p > 0$

$$\mu(H_p) = 0.$$

Roughly this result means that $C^\infty(\Omega)$-smooth path integral field sample configurations on the path probability space $(L^2(\Omega), d_A\mu(\varphi))$ make a set of zero measure Proof ([6]).

By using the following representation of the charachteristic function of the sub-set H_p

$$\chi_{H_p}(\varphi) = \lim_{\alpha\to 0}\left\{\exp\left[-\frac{\alpha}{2}\sum_{n=0}^{\infty} c_n^2 n^{2p}\right]\right\}, \tag{B-1}$$

we have the identity ($\varphi \in L^2(\Omega); \varphi \overset{L^2(\Omega)}{=} \sum c_n\varphi_n$)

$$\begin{aligned} \mu(H_p) = \lim_{\alpha=0} \int_{R^N} dj_1 \dots dj_N \\ \times \left(\int d_A\mu(\varphi)\exp\left(i\sum_{n=0}^{N} j_n c_n\right)\right) \\ \times \exp\left(-\frac{1}{2\alpha}\sum_{n=0}^{\infty}\frac{j_n^2}{n^{2p}}\right) \\ \times \left[\prod_{\mu=1}^{N}(2\pi\alpha n^{2p})^{-\frac{1}{2}}\right] \end{aligned} \tag{B-2}$$

A firect evaluation of the cylinder path integration on the right-hand side of eq. (B-4) give us the following outcome

$$\begin{aligned} \int d_A\mu(\varphi)\exp\left(i\sum_{\mu=0}^{\infty} j_n c_n\right) \\ = (2\pi)^{N/2}\exp\left\{-\frac{1}{2}\sum_{\mu=0}^{\infty}\frac{j_n^2}{\lambda_n}\right\}. \end{aligned} \tag{B-3}$$

As a consequence we have the final result on the measure of the "Sobolev Spaces" H_p

$$\mu(H_p) = \lim_{\alpha\to 0}\lim_{N\to\infty}\left\{(2\pi^{N/2}\left(\prod_{\mu=1}^{N}(2\pi\alpha\mu^{2p})^{-\frac{1}{2}}\right)\right.$$
$$\left.\times\left(\prod_{n=1}^{N}\left(\lambda_n+\frac{1}{\alpha n^{2p}}\right)^{-\frac{1}{2}}\right)\right\}$$
$$\leq \lim_{\alpha\to 0}\left\{\lim_{N\to\infty}\exp\left(-\frac{1}{2}\alpha\sum_{\mu=1}^{N}\lambda_n n^{2p}\right)\right\}$$
$$= \begin{cases} 0 & \text{if } p>0 \\ 1 & \text{if } p=0 \end{cases} \quad \text{(Minlos-Bochner Theorem)} \tag{B-4}$$

where we have used the straightforward identity

$$\prod^{N}\left(\frac{1}{1+\alpha_n}\right)^{1/2} \leq \left(\frac{1}{1+\sum^{N}\alpha_n}\right)^{1/2} \leq \exp\left\{-\frac{1}{2}\sum^{N}\alpha_N\right\}.$$

□

As it is usual to expect that $C^\infty(\Omega)\subset\bigcap_{p=0}^{\infty}H^p(\Omega)$ (where $H^p(\Omega)$ denotes the usual function Sobolev Spaces on Ω), the theorem of this appendix as expressing the fact that differentiable sample on non enough sufficiently regularized euclidean path integrals makes a set of zero measure. And classical smooth field configurations being useful only in the realm μ of formal saddle-point (WKB) path integral evaluations. So one can not manipulate path integral integrands with College Calculus rules.

For the less restringent condition of path integral sample continuity, one has to use our generalization of the Wiener theorem eq. (7.9).

In the general case of non-Gaussian cylindrical measures, one should imposes the bound restriction below, as a reasonable thechnical condition

$$\sup|Z(j)| \leq C\left[\exp\left\{-\frac{1}{2}\langle j, Aj\rangle\right\}\right] \tag{B-5}$$

for some positive definite trace class positive operator $A\in\oint_1^{+}(L^2(\Omega))$ and $C>0$, in order to obtain the validity of our theorem – Appendix B.

7.8 Appendix C

In this somewhat pedagogical appendix, we intend to presente a formal operational functional calculus to write the cylindrical Fourier Transforms as an inversible operation. We present such formal results in order to highlight the necessity of a clean distribution theory in Hilbert Spaces, get to be developed ([7]).

So let us consider a trace class, inversible and strictly positive operator A^{-1} acting on a separable Hilbert Space H.

Let $f(x) \in L^1(d_A\mu(x)(x), H) \cap L^\infty(d_A\mu(x), H)$. Since the function $\exp i\langle x, k\rangle$, for a $k \in H$ fixed is bounded, the L^1-Hilbert Space Fourier Transform is well defined

$$\hat{F}(k) = \int_H f(x) \exp i\langle x, k\rangle d_A\mu(x). \tag{C-1}$$

In the Physicist's operational notation for the cylindrical measure with $D^F[x]$ denoting the Feynman Formal (when H realized as some $L^2(\Omega)$)

$$d_A\mu(x) = \overset{+\frac{1}{2}}{\det}(A) \exp(-\frac{1}{2}(x, Ax)_H) D^F[x] \tag{C-2}$$

It still to be an open problem in Analysis in Infinite Dimensions or and Hida calculus to obtain an rigorous mathematical inversion formula for eq. (C-1). However, it is fully possible to write an inversion formula for eq. (C-1) in a more larger vectorial space: the called algebraic dual space of H i.e: H^{alg}. Let us sketchy for completeness such result of ours. As a first step one introduces an one-parameter ε ($\varepsilon \in [0,1]$) family of auxiliary inversible operators $\varepsilon^{-2}\mathbb{C}(\varepsilon)$ on $\mathcal{L}_{\text{bounded}}(H, H)$ such that $\mathbb{C}(0) = \mathbf{1}$. Let us now consider the now (perturbed) family of operator $B(\varepsilon) = (A + \varepsilon^{-2}\mathbb{C}(\varepsilon))^{-1}$. Note that $B(\varepsilon)$ exists for ε small enough and $B^{-1}(\varepsilon) = A + \varepsilon^2(\mathbb{C}(\varepsilon))$ do not belong to the trace class althought being positive definite. But even in such situation one can define a cylindrical measure in the more ample space H^{alg} through the positive-definite charachteristic functional associated to the operator $B^{-1}(\varepsilon)$

$$Z_B[x] = \int_{H^{alg}} d_{B(\varepsilon)}\nu(X) e^{iX(x)} = \exp\left\{-\frac{1}{2}\langle x, (B(\varepsilon))^{-1}x\rangle_H\right\}. \tag{C-3}$$

Let us choose our interpolating family of bounded, strictly positive operators $\mathbb{C}(\varepsilon)$ such that for $\varepsilon > 0$, $\det^{-\frac{1}{2}}(\mathbb{C}(\varepsilon)) < \infty$. We now define the following (continuous) linear functional on H for each $\varepsilon > 0$, under the non-proved hypothesis that H is a sub-set of H^{alg} of non zero measure

$$I^{(\varepsilon)}(x, [\hat{F}]) = \overset{-\frac{1}{2}}{\det}(\frac{1}{\varepsilon^2}\mathbb{C}(\varepsilon)) \left\{\int_{H^{alg}\cap H} \hat{F}(k) e^{-ik(x)} d_{B(\varepsilon)}\nu(k)\right\}. \tag{C-4}$$

Now if one substitutes eq. (C-4) into eq. (C-1) and by applying the Fubbini theorem to the Product Measure Space $(H \times H^{alg}; d_A\mu \otimes d_{B(\varepsilon)}\nu)$, one obtains the result

$$I^{(\varepsilon)}(x) = \overset{-\frac{1}{2}}{\det}(\frac{1}{\varepsilon^2}\mathbb{C}(\varepsilon))$$
$$\times \left\{ \int_{H^{alg}} \left[\int_H f(z) e^{i\langle z,k\rangle} d_A\mu(z) \right] e^{-i\langle x,k\rangle} d_{B(\varepsilon)}\nu(k) = \overset{-\frac{1}{2}}{\det}(\varepsilon^{-2}\mathbb{C}(\varepsilon)) \right.$$
$$\left. \times \left\{ \int_{H^{alg}} \int_H f(z) e^{i\langle z-x,k\rangle} (d_A\mu(z) d_{B(\varepsilon)}\nu(k)) \right\} \right.$$
$$= \overset{-\frac{1}{2}}{\det}(\varepsilon^{-2}(I(\varepsilon))) \times \left\{ \int_H f(z) Z_{B(\varepsilon)}(z-x) d_A\mu(z) \right\}$$
$$= \int_H f(z) e^{-\frac{1}{2}(z-x,A(z-x))_H}$$
$$\times \left[\overset{-\frac{1}{2}}{\det}(\varepsilon^{-2}\mathbb{C}(\varepsilon)) \exp\left(-\frac{1}{2\varepsilon^2} \langle z-x; \mathbb{C}(\varepsilon)(z-x)\rangle \right)_H \right] d_A\mu(z). \tag{C-5}$$

At this point we take from the Hida Calculus, the formal definition of the Dirac delta functional on Hilbert Spaces

$$\lim_{\varepsilon\to 0^+} \overset{-\frac{1}{2}}{\det}(\varepsilon^{-2}\mathbb{C}(\varepsilon)) \exp\left[-\frac{1}{2\varepsilon^2} \langle (z-x), \mathbb{C}(\varepsilon)|z-x|\rangle_H \right]$$
$$= \delta_H^{(F)}(z-x) \quad \text{on} \quad S'(H). \tag{C-6}$$

As a consequence, one has the operational result

$$I(x) := \lim_{\varepsilon\to 0^+} I^{(\varepsilon)}(x)$$
$$= \int_H f(z) \exp\left(-\frac{1}{2} \langle z-x, A(z-x)\rangle_H \right)$$
$$\times \delta_H^{(F)}(z-x) d_A\mu(z)$$
$$:= f(x) e^{-\frac{1}{2}\langle x,Ax\rangle} \overset{+\frac{1}{2}}{\det}(A). \tag{C-7}$$

As a consequence we have the operational formulae for Fourier Transforms in separable Hilbert Spaces.

If one has the usual cylindrical Hilbert Space Transform

$$\hat{F}(k) := \int_H f(x) \exp(i\langle x,k\rangle_H) d_A\mu(x) \tag{C-8}$$

then formally, one has the "inversion" formula on the algebraic dual of H

$$f(x) = e^{+\frac{1}{2}\langle x, Ax\rangle} \overset{-\frac{1}{2}}{\det}(A) \times \lim_{\varepsilon>0} I^{(\varepsilon)}(x, [\hat{F}]). \tag{C-9}$$

Anyway the necessity of using mathematically rigorous infinite-dimensional Fourier Transforms on Tempered Schwartz Distributions has not appeared fully yet on mathematical physics metrods. However, on light of the results presented in this paper, the time for such endoavours may be approaching ([7]).

Acknowledgments: Thanks to professor D. Pickrell of Mathematics Department of University of Arizona for discussions on $P(\phi)_2$ Field Theories on Riemman Surfaces (ref[8]).

7.9 References

[1] J. Glimm and A. Jaffe, Quantum Physics, Springer, New Yorkl, NY, USA, 2nd edition, 1987.
- B. Simon, The $P(\phi)_2$ Euclidean (Quantum) Field Theory, Princeton University Press, Princeton, NJ, USA, 1974.

[2] Luiz C.L. Botelho, Some Comments onj Rigorous Finite-Volume Euclidean Quantum Field Path Integrals in the Analytical Regularization Scheme – Hindawi Publishing Corporation, Advances in Mathematical Physics, vol 2011, Article ID 257916, DOI: 10.1155/2011/257916.

[3] Luiz C.L. Botelho, "A simple renormalization scheme in random surface theory", Modern Physics Letters B, vol 13, No. 6–7, pp. 203–207, 1999.

[4] Green, M.R., Schwarz, J.L., Witten, E., Superstring Theory, Cambridge Monographs on Mathematical Physics, vol 182, CUP, Cambridge (1996).

[5] B. Klaiber, in Lectures in Theoretical Physics: Quantum Theory and Statistical Theory, edited by A. O. Barut. Gordon and Breach, New York, 1960, vol XA, pp. 141–176.

[6] Luiz C.L. Botelho, A note an Feynman Kac path integral representations for scalar wave motions, Random Operators and Stochastic Equations (print), v. 21, pp. 271–292, (2013).

- Luiz C.L. Botelho, Semi-linear diffusion in R^D and in Hilbert Spaces, a Feynman-Wiener path integral study, Random Oper. Stoch. Equ-19 (2011), Issue 4, pages 361–386, DOI 10.1515/Rose 2011.020.

- Luiz C.L. Botelho, A method of integration for wave equation and some applications to wave physics, Random Oper. Stoch. Equ-18 (2010), No. 4, 301–325.

- Luiz C.L. Botelho, Non-Linear Diffusion and Wave Damped Propagation: Weak Solutions and Statistical Turbulence Behavior, Journal of Advanced Mathematics and Applications, vol 3, 1–11, (2014).

[7] Luiz C.L. Botelho, Lecture Notes in Applied Differential Equations of Mathematical Physics World Scientific, (2008), Singapore ISBN: 10981-281-457-4.

[8] Pickrell, Doug, $P(\phi)_2$ Quantum Field Theories and Segal's Axioms. Commun. Math. Phys. 280, 403–425, (2008).

[9] Luiz C.L. Botelho, On the rigorous ergodic theorem for a class of non-linear Klein Gordon wave propagations, Random Oper. Stoch. Equ. (March 2015), vol 23, Issue 1 DOI:10.1515/rose-2014-0029.

Chapter 8

Some comments on rigorous finite-volume euclidean quantum field path integrals in the analytical regularization scheme

Through the systematic use of the Minlos theorem on the support of cylindrical measures on R^{∞}, we produce several mathematically rigorous finite-volume euclidean path integrals in interacting euclidean quantum fields with Gaussian free measures defined by generalized powers of finite-volume Laplacean operator.

8.1 Introduction

Since the result of R.P. Feynman on representing the initial value solution of Schrodinger Equation by means of an analytically time continued integration on an infinite - dimensional space of functions, the subject of Euclidean Functional Integrals representations for Quantum Systems has became the mathematical - operational framework to analyze Quantum Phenomena and stochastic systems as showed in the previous decades of research on Theoretical Physics ([1]–[3]).

One of the most important open problem in the mathematical theory of Euclidean Functional Integrals is that related to implementation of sound mathematical approximations to these Infinite-Dimensional Integrals by means of Finite-Dimensional approximations outside of the always used [computer oriented] Space-Time Lattice approximations (see [2], [3] - chap. 9). As a first step to tackle upon the above cited problem it will be needed to characterize mathematically the Functional Domain where these Functional Integrals are defined.

The purpose of this note is to present the formulation of Euclidean Quantum Field theories as Functional Fourier Transforms by means of the Bochner-Martin-Kolmogorov theorem for Topological Vector Spaces ([4],

[5] - theorem 4.35) and suitable to define and analyze rigorously Functional Integrals by means of the well-known Minlos theorem ([5] - theorem 4.312 and [6] - part 2) which is presented in full in Appendix A.

We thus present studies on the difficult problem of defining rigorously infinite-dimensional quantum field path integrals in general finite volume space times $\Omega \subset R^\nu$ $(\nu = 2, 4, \dots)$ by means of the analytical regularization scheme ([12]).

8.2 Some rigorous finite-volume quantum field path integral in the Analytical regularization scheme

Let us thus start our analysis by considering the Gaussian measure on $L^2(R^2)$ defined by the finite volume, infrared regularized and α-power Laplacean acting on $L^2(R^N)$ as an operatorial quadratic form $(j(x) \in L^2(R^N))$ (see Appendix B)

$$\begin{aligned} Z^{(0)}_{\alpha,\varepsilon_{IR}}[j] &= \exp\left\{-\frac{1}{2}\left\langle j, (\chi_\Omega[(-\Delta)^{-\alpha} + \varepsilon^2_{IR}]^{-1}\chi_\Omega) j\right\rangle_{L^2(R^2)}\right\} \\ &\equiv \int_{L^2(R^2)} d^{(0)}_{\alpha,\varepsilon_{IR}}\mu[\varphi] \exp\left(i\left\langle j,\varphi\right\rangle_{L^2(R^2)}\right) \end{aligned} \tag{8.1-a}$$

Here χ_Ω denotes the multiplication operator defined by the characteristic function $\chi_\Omega(\alpha)$ of the compact region $\Omega \subset R^2$ and $\varepsilon_{IR} > 0$ the associated infrared cut-off.

It is worth calling the reader attention that due to the infrared regularization introduced on Eq. (8.1-a), the domain of the Gaussian measure ([4], [6]) is given by the space of square integrable functions on R^2 by the Minlos theorem of Appendix A, since for $\alpha > 1$, the operator defines a trace class operator on $L^2(R^2)$, namely

$$Tr_{\oint_1(L^2(R^2))}\{\chi_\Omega[(-\Delta)^\alpha + \varepsilon^2_{IR}]\chi_\Omega\} = \text{vol}(\Omega) \times \left[\int \frac{d^2p}{|p|^{2\alpha} + \varepsilon^2_{IR}}\right] < \infty \tag{8.1-b}$$

This is the only point of our analysis where it is needed to consider the infra-red cut off. As a consequence of the above remarks, one can analize the ultra-violet renormalization program in the following interacting model proposed by us and defined by an interaction $g_{\text{bare}}V(\varphi(x))$, with $V(x)$ being the Fourier Transformed of an integrable and essentially bounded measurable real function[1]. Note that $V(x)$ is thus a continuous real function vanishing at the infinite point.

Let us show that by defining an ultra-violet renormalized coupling constant (with a finite volume Ω cut off built in).

$$g_{\text{bare}}(\alpha) = \frac{g_{\text{ren}}}{(1-\alpha)^{1/2}}\left(2^{-\alpha}\pi^{-\frac{1}{4}}\right) \tag{8.2}$$

one can show that the interaction function

$$\exp\left\{-g_{\text{bare}}(\alpha)\int_{\Omega} d^2x\, V(\varphi(x))\right\} \tag{8.3}$$

is an integrable function on $L^1(L^2(R^2), d^{(0)}_{\alpha,\varepsilon_{IR}}\mu\,[\varphi])$ and leads to a well-defined ultra-violet functional integral in the limit of $\alpha \to 1$.

The proof is based on the following estimates.

Since almost everywhere we have the pointwise limit

$$\exp\left\{-g_{\text{bare}}(\alpha)\int d^2x\, V(\varphi(x))\right\}$$

$$\lim_{N\to\infty}\left\{\sum_{n=0}^{N}\frac{(-1)^n(g_{\text{bare}}(\alpha))^n}{n!}\int_{[-\Lambda,\Lambda]} dk_1\cdots dk_n\,\tilde{V}(k_1)\cdots\tilde{V}(k_n)\right.$$
$$\left.\times\int_{\Omega} dx_1\cdots dx_n\, e^{ik_1\varphi(x_1)}\cdots e^{ik_n\varphi(x_n)}\right\} \tag{8.4}$$

we have that the upper-bound estimate below holds true

$$\left|Z^{\alpha}_{\varepsilon_{IR}}[g_{\text{bare}}]\right| \leq \left|\sum_{n=0}^{\infty}\frac{(-1)^n(g_{\text{bare}}(\alpha))^n}{n!}\int_R dk_1\cdots dk_n\, V(k_1)\cdots\tilde{V}(k_n)\right.$$
$$\left.\int_{\Omega} dx_1\cdots dx_n\int d^{(0)}_{\alpha,\varepsilon_{IR}}\mu[\varphi](e^{i\sum_{\ell=1}^{N}k_\ell\varphi(x_\ell)})\right| \tag{8.5-a}$$

with

$$Z^{\alpha}_{\varepsilon_{IR}}[g_{\text{bare}}] = \int d^{(0)}_{\alpha,\varepsilon_{IR}}\mu[\varphi]\exp\left\{-g_{\text{bare}}(\alpha)\int_{\Omega} d^2x\, V(\varphi(x))\right\} \tag{8.5-b}$$

---

[1]It could be as well consider also a polinomial interaction of the form $V_{n,p}(x) =$ minimum of $\{(\varphi(x))^p, n\}$ with p and n psotive integers. Note that $\tilde{V}(k) \in L^1(R)\cap L^\infty(R)$ by hypothesis

$$\left|\int_{\Omega}\left[\int e^{ik\varphi(x)}\cdot\tilde{V}(k)dk\right]d^2x\right| \leq \text{vol}(\Omega)||\tilde{V}||_{L^1} < \infty$$

we have, thus, the more suitable form after realizing the d^2k_i and $d^{(0)}_{\alpha,\varepsilon_{IR}}\mu[\varphi]$ integrals respectivelly[2]

$$\left|Z^{\alpha}_{\varepsilon_{IR}=0}[g_{bare}]\right| \le \sum_{n=0}^{\infty} \frac{(g_{bare}(\alpha))^n}{n!} \left(||\tilde{V}||_{L^\infty(R)}\right)^n \left|\int_{R^2} dx_1 \cdots dx_n \det{}^{-\frac{1}{2}} \left[G^{(N)}_{\alpha}(x_i,x_j)\right]_{\substack{1\le i\le N\\ 1\le j\le N}}\right| \tag{8.6}$$

Here $[G^{(N)}_{\alpha}(x_i,x_j)]_{\substack{1\le i\le N\\ 1\le j\le N}}$ denotes the $N\times N$ symmetric matrix with the (i,j) entry given by the positive Green-function of the α-Laplacean (without the infra-red cut off here!).

$$G_\alpha(x_i,x_j) = |x_i - x_j|^{2(\alpha-1)} \frac{\Gamma(1-\alpha)}{\Gamma(\alpha)} \frac{1}{2\pi} 2^{2(1-\alpha)} \tag{8.7}$$

At this point, we call the reader attention that we have the formulae on the asymptotic behavior for $\alpha \to 1$ and $\alpha < 1$ (see ref. [12] - Appendix A).

$$\left\{\lim_{\substack{\alpha\to 1\\ \alpha>1}} \det{}^{-\frac{1}{2}}[G^{(N)}_{\alpha}(x_i,x_j)]\right\} \sim e^{N\pi i\alpha} \times (\pi^{\frac{N}{4}} \cdot 2^{N\alpha}) \times (1-\alpha)^N \left(\frac{(+1)}{(-1)^N(N-1)}|\right) \tag{8.8}$$

After substituting eq. (8.8) into eq. (8.6) and taking into account the hypothesis of the compact support of the nonlinearity $\tilde{V}(k)$, one obtains the finite bound for any value $g_{rem} > 0$, without the finite volume cut off and producing a proof for the convergence of the perturbative expansion in terms of the renormalized coupling constant for the model

$$\begin{aligned}\lim_{\alpha\to 1}\left|Z^{\alpha}_{\varepsilon_{IR}=0}[g_{bare}(\alpha)]\right| &\le \sum_{n=0}^{\infty} \frac{(\|\tilde{V}\|_{L^\infty(R)})^n}{n!} \left(\frac{g_{ren}}{(1-\alpha)^{\frac{1}{2}}}\right)^n \\ &\quad \times (1-\alpha)^{n/2} \times (\mathrm{vol}(\Omega))^n \\ &\le \exp\{g_{ren}||V||_{L^\infty(R)}\mathrm{vol}(\Omega)\} < \infty \end{aligned} \tag{8.9}$$

Another important rigorously defined functional integral is to consider the following α-power Klein Gordon operator on Euclidean space-time

$$\mathcal{L}^{-1}_{\Omega} = \left(\frac{\chi_\Omega (2\pi)^{\nu/2}}{\mathrm{vol}(\Omega)}\right) [(-\Delta)^\alpha + m^2]^{-1} \times \left(\frac{\chi_\Omega (2\pi)^{\nu/2}}{\mathrm{vol}(\Omega)}\right) \tag{8.10}$$

[2]Note that:

$$\int_\Omega d^2x_1 \dots d^2x_n e^{-\frac{1}{2}\sum_{i,j}^N k_i k_j \overbrace{G_\alpha(x_i,x_j)}^{\ge 0}} \le \int_{R^2} d^2x_1 \dots d^2x_n e^{-\frac{1}{2}\sum_{i,j}^N k_i k_j \overbrace{G_\alpha(x_i,x_j)}^{\ge 0}}$$

with m^2 a positive "mass" parameters.

Let us note that $\mathcal{L}^{-1}$ is an operator of trace class on $L^2(R^\nu)$ if and only if the result below holds true

$$Tr_{L^2(R^\nu)}(\mathcal{L}^{-1}) = \int d^\nu k \frac{1}{k^{2\alpha}+m^2} = \bar{C}(\nu)\, m^{(\frac{\nu}{\alpha}-2)} \times \left\{ \frac{\pi}{2\alpha} \mathrm{cosec} \frac{\nu\pi}{2\alpha} \right\} < \infty \tag{8.11}$$

namely if

$$\alpha > \frac{\nu}{2} \tag{8.12}$$

In this case, let us consider the double functional integral with functional domain $L^2(R^\nu)$

$$\begin{aligned} Z[j,k] = &\int d_G^{(0)}\, \beta[v(x)] \\ &\times \int d_{\mathcal{L}\Omega}^{(0)} \mu[\varphi] \\ &\times \exp\left\{ i \int d^\nu x \, (j(x)\, \varphi(x) + k(x)\, v(x)) \right\} \end{aligned} \tag{8.13}$$

where the Gaussian functional integral on the fields $V(x)$ has a Gaussian generating functional defined by a $\oint_1$-integral operator with a positive-definite kernel $g(|x-y|)$, namely

$$\begin{aligned} Z^{(0)}[k] &= \int d_G^{(0)} \beta[v(x)] \, \exp\left\{ i \int d^\nu x \, k(x) v(x) \right\} \\ &= \exp\left\{ -\frac{1}{2} \int_\Omega d^\nu x \int_\Omega d^\nu y \, (k(x)\, g(|x-y|)\, k(y)) \right\} \end{aligned} \tag{8.14}$$

By a simple direct application of the Fubbini-Tonelli theorem on the exchange of the integration order on eq. (8.13), lead us to the effective $\lambda\varphi^4$ - like well-defined functional integral representation

$$\begin{aligned} Z_{\text{eff}}[j] = &\int d_{\mathcal{L}}^{(0)} \mu[\varphi][\varphi(x)] \\ &\exp\left\{ -\frac{1}{2} \int_\Omega d^\nu x d^\nu y \, |\varphi(x)|^2 \, g(|x-y|) \, |\varphi(y)|^2 \right\} \\ &\times \exp\left\{ i \int_\Omega d^\nu x \, j(x) \varphi(x) \right\} \end{aligned} \tag{8.15}$$

Note that if one introduces from the begining a bare mass parameters m^2_{bare} depending on the parameters α, but such that it always satisfies eq. (8.11) one should obtains again eq. (8.15) as a well-defined measure on $L^2(R^\nu)$. Of course that the usual pure Laplacean limit of

$\alpha \to 1$ on eq. (8.10), will needed a renormalization of this mass parameters $(\lim_{\alpha\to 1} m^2_{bare}(\alpha) = +\infty!)$ as much as it has been done in previous studies.

Let us continue our examples by showing again the usefulness of the precise determination of the functional - distributional structure of the domain of the functional integrals in order to construct rigorously these path integrals without complicated lattice limit procedures ([2]).

Let us consider a general R^ν Gaussian measure defined by the Generating functional on $S(R^\nu)$ defined by the α-power of the Laplacean operator $-\Delta$ acting on $S(R^\nu)$ with a small infrared regularization mass parameter μ^2 as considered in eq. (8.1-a)

$$
\begin{aligned}
Z_{(0)}[j] &= \exp\left\{-\frac{1}{2}\left\langle j, ((-\Delta)^\alpha + \mu_0^2)^{-1} j\right\rangle_{L^2(R^\nu)}\right\} \\
&= \int_{E^{alg}(S(R^\nu))} d_\alpha^{(0)}\,\mu[\varphi]\,\exp(i\,\varphi(j))
\end{aligned}
\tag{8.16}
$$

An explicit expression in momentum space for the Green function of the α-power of $(-\Delta)^\alpha + \mu_0^2$ given by

$$
((-\Delta)^{+\alpha} + \mu_0^2)^{-1}(x-y) = \int \frac{d^\nu k}{(2\pi)^\nu}\, e^{ik(x-y)} \left(\frac{1}{k^{2\alpha} + \mu_0^2}\right) \tag{8.17}
$$

Here $\bar{C}(\nu)$ is a ν-dependent (finite for ν-values!) normalization factor.

Let us suppose that there is a range of α-power values that can be choosen in such way that one satisfies the constraint below

$$
\int_{E^{alg}(S(R^\nu))} d_\alpha^{(0)}\,\mu[\varphi](\|\varphi\|_{L^{2j}(R^\nu)})^{2j} < \infty \tag{8.18}
$$

with $j = 1, 2, \cdots, N$ and for a given fixed integer N, the highest power of our polinomial field interaction. Or equivalently, after realizing the φ-Gaussian functional integration, with a space-time cutt off volume Ω on the interaction to be analyzed on eq. (8.16)

$$
\begin{aligned}
\left|\int_\Omega d^\nu x[(-\Delta)^\alpha + \mu_0^2]^{-j}(x,x)\right| &\le \left|\mathrm{vol}(\Omega) \times \left(\int \frac{d^\nu k}{k^{2\alpha} + \mu_0^2}\right)^j\right| \\
&= \left|\left[C_\nu(\mu_0)^{(\frac{\nu}{\alpha}-2)} \times \left(\frac{\pi}{2\alpha}\mathrm{cosec}\frac{\nu\pi}{2\alpha}\right)\right]^j\right| < \infty
\end{aligned}
\tag{8.19}
$$

For $\alpha > \frac{\nu}{2}$, one can see by the Minlos theorem that the measure support of the Gaussian measure eq. (8.16) will be given by the intersection Banach space of measurable Lebesgue functions on R^ν instead of the previous one $E^{alg}(S(R^\nu))$ ([4]–[6]).

$$\mathcal{L}_{2N}(R^\nu) = \bigcap_{j=1}^{N} (L^{2j}(R^\nu)) \tag{8.20}$$

In this case, one obtains that the finite - volume $p(\varphi)_2$ interactions

$$\exp\left\{ -\sum_{j=1}^{N} \lambda_{2j} \int_\Omega (\varphi^2(x))^j \, dx \right\} \leq 1 \tag{8.21}$$

is mathematically well-defined as the usual pointwise product of measurable functions and for positive coupling constant values $\lambda_{2j} \geq 0$. As a consequence, we have a measurable functional on $L^1(\mathcal{L}_{2N}(R^\nu);\, d_\alpha^{(0)}\, \mu[\varphi])$ (since it is bounded by the function 1). Thus, it makes sense to consider mathematically the well-defined path - integral on the full space R^ν with those values of the power α satisfying the contraint eq. (8.17).

$$Z[j] = \int_{\mathcal{L}_{2N}(R^\nu)} d_\alpha^{(0)}\, \mu[\varphi] \, \exp\left\{ -\sum_{j=1}^{N} \lambda_{2j} \int_\Omega \varphi^{2j}(x) dx \right\} \times \exp(i \int_{R^\nu} j(x)\varphi(x)) \tag{8.22}$$

Finally, let us consider an interacting field theory in a compact space-time $\Omega \subset R^\nu$ defined by an iteger even power $2n$ of the Laplacean operator with Dirichlet Boundary conditions as the free Gaussian kinetic action, namely

$$\begin{aligned} Z^{(0)}[j] &= \exp\left\{ -\frac{1}{2} \left\langle j, (-\Delta)^{2n} j \right\rangle_{L^2(\Omega)} \right\} \\ &= \int_{W_2^n(\Omega)} d_{(2n)}^{(0)}\, \mu[\varphi] \, \exp(i\langle j, \varphi\rangle_{L^2(\Omega)}) \end{aligned} \tag{8.23}$$

here $\varphi \in W_2^n(\Omega)$ - the Sobolev space of order n which is the functional domain of the cylindrical Fourier Transform measure of the Generating functional $Z^{(0)}[j]$, a continuous bilinear positive form on $W_2^{-n}(\Omega)$ (the topological dual of $W_2^n(\Omega)$) ([4]–[6]).

By a straightforward application of the well-known Sobolev immersion theorem, we have that for the case of

$$n - k > \frac{\nu}{2} \tag{8.24}$$

including k a real number the functional Sobolev space $W_2^n(\Omega)$ is contained in the continuously fractional differentiable space of functions $C^k(\Omega)$. As a consequence, the domain of the Bosonic functional integral can be further reduced to $C^k(\Omega)$ in the situation of eq. (8.24)

$$Z^{(0)}[j] = \int_{C^k(\Omega)} d^{(0)}_{(2n)}\,\mu[\varphi]\,\exp(i\langle j,\varphi\rangle_{L^2(\Omega)}) \tag{8.25}$$

That is our new result generalizing the Wiener theorem on Brownian paths in the case of $n = 1$, $k = \frac{1}{2}$ and $\nu = 1$

Since the bosonic functional domain on eq. (8.25) is formed by real functions and not distributions, we can see straightforwardly that any interaction of the form

$$\exp\left\{-g\int_\Omega F(\varphi(x))d^\nu x\right\} \tag{8.26}$$

with the non-linearity $F(x)$ denoting a lower bounded real function ($\gamma > 0$)

$$F(x) \geq -\gamma \tag{8.27}$$

is well-defined and is integrable function on the functional space $(C^k(\Omega),\, d^{(0)}_{(2n)}\,\mu[\varphi])$ by a direct application of the Lebesgue theorem

$$\left|\exp\left\{-g\int_\Omega F(\varphi(x))\,d^\nu x\right\}\right| \leq \exp\{+g\gamma\} \tag{8.28}$$

At this point we make a subtle mathematical remark that the infinite volume limit of eq. (8.25)–eq. (8.26) is very difficult, since one looses the Garding - Poincar inequalite at this limit for those elliptic operators and, thus, the very important Sobolev theorem. The probable correct procedure to consider the thermodynamic limit in our Bosonic path integrals is to consider solely a volume cut off on the interaction term Gaussian action as in eq. (8.22) and there search for vol$(\Omega) \to \infty$ ([7]–[10]).

As a last remark related to eq. (8.23) one can see that a kind of "fishnet" exponential generating functional

$$Z^{(0)}[j] = \exp\left\{-\frac{1}{2}\left\langle j, \exp\{-\alpha\Delta\}j\right\rangle_{L^2(\Omega)}\right\} \tag{8.29}$$

has a Fourier transformed functional integral representation defined on the space of the infinitely differentiable functions $C^\infty(\Omega)$, which physically means that all field configurations making the domain of such path integral has a strong behavior like purely nice smooth classical field configurations.

As a last important point of this note, we present an important result on the geometrical characterization of massive free field on an Euclidean Space-Time ([10]).

Firstly we announce a slightly improved version of the usual Minlos Theorem ([4]).

Theorem 3. Let E be a nuclear space of tests functions and $d\mu$ a given σ-measure on its topologic dual with the strong topology. Let $\left\langle \, , \right\rangle_0$ be an inner product in E, inducing a Hilbertian structure on $\mathcal{H}_0 = \overline{(E, \left\langle \, , \right\rangle_0)}$, after its topological completation.

We suppose the following:

a) There is a continuous positive definite functional in $\mathcal{H}_0$, $Z(j)$, with an associated cylindrical measure $d\mu$.

b) There is a Hilbert-Schmidt operator $T\colon \mathcal{H}_0 \to \mathcal{H}_0$; invertible, such that $E \subset$ Range (T), $T^{-1}(E)$ is dense in $\mathcal{H}_0$ and $T^{-1}\colon \mathcal{H}_0 \to \mathcal{H}_0$ is continuous.

We have thus, that the support of the measure satisfies the relationship

$$\text{support } d\mu \subseteq (T^{-1})^*(\mathcal{H}_0) \subset E^* \tag{8.30}$$

At this point we give a non-trivial application of ours of the above cited Theorem 3.

Let us consider an differential inversible operator $\mathcal{L}\colon S'(R^N) \to S(R)$, together with an positive inversible self-adjoint elliptic operator $P\colon D(P) \subset L^2(R^N) \to L^2(R^N)$. Let H_α be the following Hilbert space

$$H_\alpha = \left\{ \overline{S(R^N), \left\langle P^\alpha \varphi, P^\alpha \varphi \right\rangle_{L^2(R^N)} = \left\langle \, , \right\rangle_\alpha}, \text{ for } \alpha \text{ a real number} \right\}. \tag{8.31}$$

We can see that for $\alpha > 0$, the operators below

$$\begin{aligned} P^{-\alpha}\colon L^2(R^N) &\to \mathcal{H}_{+\alpha} \\ \varphi &\to (P^{-\alpha}\varphi) \end{aligned} \tag{8.32}$$

$$\begin{aligned} P^{\alpha}\colon \mathcal{H}_{+\alpha} &\to L^2(R^N) \\ \varphi &\to (P^{\alpha}\varphi) \end{aligned} \tag{8.33}$$

are isometries among the following sub-spaces

$$\overline{D(P^{-\alpha}), \left\langle \, , \right\rangle_{L^2}}) \text{ and } H_{+\alpha}$$

since

$$\left\langle P^{-\alpha}\varphi, P^{-\alpha}\varphi \right\rangle_{\mathcal{H}_{+\alpha}} = \left\langle P^\alpha P^{-\alpha}\varphi, P^\alpha P^{-\alpha}\varphi \right\rangle_{L^2(R^N)} = \left\langle \varphi, \varphi \right\rangle_{L^2(R^N)} \tag{8.34}$$

and

$$\left\langle P^\alpha f, P^\alpha f \right\rangle_{L^2(R^N)} = \left\langle f, f \right\rangle_{H_{+\alpha}} \tag{8.35}$$

If one considers T a given Hilbert-Schmidt operator on H_α, the composite operator $T_0 = P^\alpha T P^{-\alpha}$ is an operator with domain being $D(P^{-\alpha})$ and its image being the Range (P^α). T_0 is clearly an invertible operator and $S(R^N) \subset$ Range (T) means that the equation $(TP^{-\alpha})(\varphi) = f$ has always a non-zero solution in $D(P^{-\alpha})$ for any given $f \in S(R^N)$. Note that the condition that $T^{-1}(f)$ be a dense subset on Range $(P^{-\alpha})$ means that

$$\left\langle T^{-1} f, P^{-\alpha}\varphi \right\rangle_{L^2(R^N)} = 0 \tag{8.36}$$

has as unique solution the trivial solution $f \equiv 0$.

Let us suppose too that T^{-1}: $S(R^N) \to H_\alpha$ be a continuous application and the bilinear term $(\mathcal{L}^{-1}(j))(j)$ be a continuous application in the Hilbert spaces $H_{+\alpha} \supset S(R^N)$, namely: if $j_n \xrightarrow{L^2} j$, then $\mathcal{L}^{-1}$: $P^{-\alpha} j_n \xrightarrow{L^2} \mathcal{L}^{-1} P^{-\alpha} j$, for $\{j_n\}_{n \in \mathbb{Z}}$ and $j_n \in S(R^N)$.

By a direct application of the Minlos Theorem, we have the result

$$Z(j) = \exp\left\{ -\frac{1}{2} [\mathcal{L}^{-1}(j)(j)] \right\} = \int_{(T^{-1})^* H_\alpha} d\mu(T) \exp(iT(j)) \tag{8.37}$$

Here the topological space support is given by

$$\begin{aligned}(T^{-1})^* \mathcal{H}_\alpha &= \left[\left(P^{-\alpha} T_0 P^\alpha \right)^{-1} \right]^* \left(\overline{(P^\alpha(S(R^N)))} \right) \\ &= \left[(P^\alpha)^* (T_0^{-1})^* (P^{-\alpha)})^* \right] P^\alpha (S(R^N)) \\ &= P^\alpha T_0^{-1} (L^2(R^N))\end{aligned} \tag{8.38}$$

In the important case of $\mathcal{L} = (-\Delta + m^2)$: $S'(R^N) \to S(R^N)$ and $T_0 T_0^* = \frac{\chi_\Omega (2\pi)^{N/2}}{\mathrm{vol}(\Omega)} (-\Delta + m^2)^{-2\beta} \times \chi_\Omega \in \oint_1 (L^2(R^N))$ since $Tr(T_0 T_0^*) = \frac{1}{2(m^2)^\beta} \left(\frac{m^2}{1} \right)^{\frac{N}{2}} \frac{\Gamma(\frac{N}{2}) \Gamma(2\beta - \frac{N}{2})}{\Gamma(\beta)} < \infty$ for $\beta > \frac{N}{4}$ with the choice $P = (-\Delta + m^2)$, we can see that the support of the measure in the path-integral representation of the Euclidean measure field in R^N may be taken as the measurable sub-set below

$$\operatorname{supp}\{d_{(-\Delta+m^2)} u(\varphi)\} = (-\Delta + m^2)^{+\alpha} I_\Omega(x) (-\Delta + m^2)^{+\beta} (L^2(R^N)) \tag{39}$$

since $\mathcal{L}^{-1}P^{-\alpha} = (-\Delta + m^2)^{-1-\alpha}$ is always a bounded operator in $L^2(R^N)$ for $\alpha > -1$.

As a consequence each field configuration can be considered as a kind of "fractional distributional" derivative of a square integrable function as written below of the formal infinite volume $\Omega \to R^N$.

$$\varphi(x) = \left[\left(- \Delta + m^2\right)^{\frac{N}{4}+\varepsilon-1} f\right](x) \tag{8.40}$$

with a function $f(x) \in L^2(R^N)$ and any given $\varepsilon > 0$, even if originally all fields configurations entering into the path-integral were elements of the Schwartz Tempered Distribution Spaces $S'(R^N)$ certainly very "rough" mathematical objects to characterize from a rigorous geometrical point of view.

We have, thus, make a further reduction of the functional domain of the free massive Euclidean scalar field of $S'(R^N)$ to the measurable sub-set as given by eq. (8.30) denoted by $W(R^N)$

$$\exp\left\{-\frac{1}{2}[(-\Delta + m^2)^{-1}j](j)\right\} = \int_{S'(R^N)} d_{(-\Delta+m^2)}\mu(\varphi)\, e^{i\,\varphi(j)}$$

$$= \int_{W(R^N)\subset S'(R^N)} d_{(-\Delta+m^2)}\tilde{\mu}(f)\, e^{i\left\langle f,(-\Delta+m^2)^{\frac{N}{4}+\varepsilon-1} f\right\rangle_{L^2(R^N)}} \tag{8.41}$$

8.3 References

[1] B. Simon, "Functional Integration and Quantum Physics" - Academic Press, (1979).

[2] B. Simon, "The $P(\phi)_2$ Euclidean (Quantum) Field Theory"- Princeton Series in Physics, (1974).

[3] J. Glimm and A. Jaffe, "Quantum Physics" - A Functional Integral Point of View Springer Verlag, (1988).

[4] Y. Yamasaki, "Measure on Infinite Dimensional Spaces" - World Scientific Vol. 5, (1985).

[5] Xia Dao Xing, "Measure and Integration Theory on Infinite Dimensional Space" Academic Press, (1972).

[6] L. Schwartz, "Random Measure on Arbitrary Topological Space and Cylindrical Measures" - Tata Institute - Oxford University Press, (1973).

[7] K. Symanzik, J. Math., Phys., 7, 510 (1966).

[8] B. Simon, Journal of Math. Physics, vol. 12, 140 (1971).

[9] Luiz C.L. Botelho, Phys., Rev. 33D , 1195 (1986).
[10] E Nelson, Regular probability measures on function space; Annals of Mathematics (2), 69, 630 - 643, (1959).
[11] Walter Rudin, Real and Complex Analysis, second edition, Tata McGraw-Hill, Publishing Co Limited, New Delhi, (1979).
[12] Luiz C.L. Botelho - "A Simple Renormalization Scheme in Random Surface Theory" - Modern Physics Letters, 13B, n$^{\underline{os}}$ 6 and 7, 203–207, (1999).

8.4 Appendix A: Some Comments on the Support of Functional Measures in Hilbert Space

Let us comment further on the application of the Minlos Theorem in Hilbert Spaces. In this case one has a very simple proof which holds true in general Banach Spaces $(E, || \quad ||)$.

Let us thus, give a cylindrical measures $d^\infty \mu(x)$ in the algebraic dual E^{alg} of a given Banach Space E ([4]–[6]).

Let us suppose either that the function $||x||$ belongs to $L^1(E^{\text{alg}}, d^\infty \mu(x))$. Then the support of this cylindrical measures will be the Banach Space E.

The proof is the following:

Let A be a sub-set of the vectorial space E^{alg} (with the topology of pontual convergence), such that $A \subset E^c$ (so $||x| = +\infty$) (E can always be imbed as a cylindrical measurable sub-set of E^{alg} - just use a Hammel vectorial basis to see that). Let be the sets $A_n = \{x \in E^{\text{alg}} \mid ||x|| \geq n\}$. Then we have the set inclusion $A \subset \bigcap_{n=0}^{\infty} A_n$, so its measure satisfies the estimates below:

$$\begin{aligned}
\mu(A) &\leq \liminf_n \mu(A_n) \\
&= \liminf_n \mu\{x \in E^{\text{alg}} \mid ||x|| \geq n\} \\
&\leq \liminf_n \left\{ \frac{1}{n} \int_{E^{\text{alg}}} ||x|| d^\infty \mu(x) \right\} \\
&= \liminf_n \frac{||x||_{L^1(E^{\text{alg}}, d_\mu^\infty)}}{n} = 0. \qquad \text{(A-1)}
\end{aligned}$$

Leading us to the Minlos theorem that the support of the cylindrical measure in E^{alg} is reduced to the own Banach Space E.

Note that by the Minkowisky inequality for general integrals, we have that $||x||^2 \in L^1(E^{\text{alg}}, d^\infty \mu(x))$. Now it is elementary evaluation to see that if $A^{-1} \in \oint_1(\mathcal{M})$, when $E = \mathcal{M}$, a given Hilbert Space, we have that

$$\int_{\mathcal{M}^{\text{alg}}} d_A^\infty \mu(x) \cdot ||x||^2 = Tr_{\mathcal{M}}(A^{-1}) < \infty. \tag{A-2}$$

This result produces another criterium for supp $d_A^\infty \mu = \mathcal{M}$ (the Minlos Theorem), when $E = \mathcal{M}$ is a Hilbert Space.

It is easy too to see that if

$$\int_{\mathcal{M}} ||x|| d^\infty \mu(x) < \infty \tag{A-3}$$

then the Fourier-Transformed functional

$$Z(j) = \int_{\mathcal{M}} e^{i(j,x)_{\mathcal{M}}} d^\infty \mu(x) \tag{A-4}$$

is continuous in the norm topology of $\mathcal{M}$.

Otherwise, if $Z(j)$ is not continuous in the origin $0 \in \mathcal{M}$ (without loss of generality), then there is a sequence $\{j_n\} \in \mathcal{M}$ and $\delta > 0$, such that $||j_n|| \to 0$ with

$$\begin{aligned} \delta \leq |Z(j_n) - 1| &\leq \int_{\mathcal{M}} |e^{i(j_n,x)_{\mathcal{M}}} - 1| d^\infty \mu(x) \\ &\leq \int_{\mathcal{M}} |(j_n, x)| d^\infty \mu(x) \\ &\leq ||j_n|| \Big(\int_{\mathcal{M}} ||x|| d^\infty \mu(x) \Big) \to 0, \end{aligned} \tag{A-5}$$

a contradiction with $\delta > 0$.

Finally, let us consider an elliptic operator B (with inverse) from the Sobelev space $\mathcal{M}^{-2m}(\Omega)$ to $\mathcal{M}^{2m}(\Omega)$. Then by the criterium given by Eq. (A-2) if

$$Tr_{L^2(\Omega)}[(I + \Delta)^{+\frac{m}{2}} B^{-1} (I + \Delta)^{+\frac{m}{2}}] < \infty, \tag{A-6}$$

we will have that the path integral below written is well-defined for $x \in \mathcal{M}^{+2m}(\Omega)$ and $j \in \mathcal{M}^{-2m}(\Omega)$. Namely

$$\exp(-\frac{1}{2}(j, B^{-1} j)_{L^2(\Omega)}) = \int_{\mathcal{M}^{+2m}(\Omega)} d_B \mu(x) \exp(i(j, x)_{L^2(\Omega)}). \tag{A-7}$$

By the Sobolev theorem which means that the embeeded below is continuous (with $\Omega \subseteq R^\nu$ denoting a smooth domain), one can further reduce the measure support to the Hlder α continuous function in Ω if $2m - \frac{\nu}{2} > \alpha$. Namely, we have a easy proof of the famous Wiener Theorem on sample continuity of certain path integrals in Sobolev Spaces

$$\mathcal{M}^{2m}(\Omega) \subset C^\alpha(\Omega) \tag{A-8-a}$$

The above Wiener Theorem is fundamental in order to construct non-trivial examples of mathematically rigorous euclideans path integrals in spaces R^ν of higher dimensionality, since it is a trivial consequence of the Lebesgue theorem that positive continuous functions $V(x)$ generate functionals integrable in $\{\mathcal{M}^{2m}(\Omega), d_B\mu(\varphi)\}$ of the form below

$$\exp\left\{-\int_\Omega V(\varphi(x))dx\right\} \in L^1(\mathcal{M}^{2m}(\Omega), d_B\mu(\varphi)). \tag{A-8-b}$$

As a last important remark on Cylindrical Measures in Separable Hilbert Spaces, let us point at to our reader that the support of such above measures is always a σ-compact set in the norm topology of $\mathcal{M}$. In order to see such result let us consider a given dense set of $\mathcal{M}$, namely $\{x_k\}_{k\in I^+}$. Let $\{\delta_k\}_{k\in I^+}$ be a given sequence of positive real numbers with $\delta_k \to 0$. Let $\{\varepsilon_n\}$ another sequence of positive real numbers such that $\sum_{n=1}^\infty \varepsilon_n < \varepsilon$. Now it is straightforward to see that $\mathcal{M} \subset \bigcup_{h=1}^\infty \overline{B(x_k,\delta_k)} \subset \mathcal{M}$ and thus $\limsup \mu\{\overline{\bigcup_{k=1}^n B(x_k,\delta_k)}\} = \mu(\mathcal{M}) = 1$. As a consequence, for each n, there is a k_n, such that $\mu\left(\bigcup_{k=1}^{k_n} \overline{B(x_k,\delta_k)}\right) \geq 1-\varepsilon$.

Now the sets $K_\mu = \bigcap_{n=1}^\infty \left[\bigcup_{k=1}^{k_n} \overline{B(x_k,\delta_k)}\right]$ are closed and totally bounded, so they are compact sets in $\mathcal{M}$ with $\mu(\mathcal{M}) \geq 1-\varepsilon$. Let is now choose $\varepsilon = \frac{1}{n}$ and the associated compact sets $\{K_{n,\mu}\}$. Let us further consider the compact sets $\hat{K}_{n,\mu} = \bigcup_{\ell=1}^n K_{\ell,\mu}$. We have that $\hat{K}_{n,\mu} \subseteq \hat{K}_{n+1,\mu}$, for any n and $\limsup \mu(\hat{K}_{n,\mu}) = 1$. So, supp $d\mu = \bigcup_{n=1}^\infty \hat{K}_{n,\mu}$, a σ-compact set of $\mathcal{M}$.

We consider now a enumerable family of cylindrical measures $\{d\mu_n\}$ in $\mathcal{M}$ satisfying the chain inclusion relationship for any $n \in I^+$

$$\text{supp } d\mu_n \subseteq \text{supp } d\mu_{n+1}.$$

Now it is straightforward to see that the compact sets $\{\hat{K}_n^{(n)}\}$, where supp $d\mu_m = \bigcup_{n=1}^\infty \hat{K}_n^{(m)}$, is such that supp $\{d\mu_m\} \subseteq \bigcup_{n=1}^\infty \hat{K}_n^{(n)}$, for any $m \in I^+$.

Let us consider the family of functionals induced by the restriction of this sequence of measures in any compact $\hat{K}_n^{(n)}$. Namely

$$\mu_n \to L_n^{(n)}(f) = \int_{\hat{K}_n^{(n)}} f(x)\cdot d\mu_p(x). \tag{A-8-c}$$

Here $f \in C_b(\hat{K}_n^{(n)})$. Note that all the above functionals in $\bigcup_{n=1}^\infty C_b(\hat{K}_n^{(n)})$ are bounded by 1. By the Alaoglu-Bourbaki theorem they form a compact set in the weak star topology of $\left(\bigcup_{n=1}^\infty C_b(\hat{K}_n^{(n)})\right)^*$, so there is a subsequence (or better the whole sequence) converging to a unique cylindrical measure $\bar{\mu}(x)$. Namely

$$\lim_{n\to\infty} \int_{\mathcal{M}} f(x) d\mu_n(x) = \int_{\mathcal{M}} f(x) d\bar{\mu}(x) \tag{A-8-d}$$

for any $f \in \bigcup_{n=1}^{\infty} C_b(\hat{K}_n^{(n)})$.

8.5 Appendix B

A straightforward calculation give the following expression for the trace of the integral operator on Eq. (A-1)

$$\mathrm{Tr}_{\oint_1}\{\chi_\Omega[(-\Delta)^\alpha + \varepsilon_{IR}^2]^{-1}\chi_\Omega\} = \int_{-\infty}^{+\infty} \left\{ \int_{-\infty}^{+\infty} \left[\frac{\widetilde{\chi}_\Omega(k-p)\widetilde{\chi}_\Omega(p-k)}{(p^{2\alpha}+\varepsilon_{IR}^2)} \right] !dp \right\} dk, \tag{B-1}$$

where

$$\widetilde{\chi}_\Omega(p) = \frac{1}{2\pi} \left\{ \int_{-\infty}^{+\infty} \chi_\Omega(x) e^{ipx}\, d^2x \right\}. \tag{B-2}$$

Since by Parservals theorem

$$\int_{-\infty}^{+\infty} dk\, |\widetilde{\chi}_\Omega(k-p)|^2 d^2k = \int_{-\infty}^{+\infty} |\chi_\Omega(x)|^2 d^2x = \mathrm{vol}(\Omega) \tag{B-3}$$

one has the validity of the result written on Eq. (B-1).

Chapter 9

On the Rigorous Ergodic Theorem for a Class of Non-Linear Klein Gordon Wave Propagations

We present a complete study of the Ergodic Theorem for the difficult problem of Non-Linear Klein Gordon Classical Wave Propagations through Cylindrical Measures, rigorous mathematical Path Integrals and the famous Ruelle-Amrein-Geogerscu-Enss (R.A.G.E.) theorem on the caracterization of continuous spectrum of self-adjoint operators

9.1 Introduction

One of the most important phenomenon in numerical studies of the non-linear wave motion especially in the two - dimensional case - is the existence of overwhelming majority of wave motions that wandering over all possible system's phase space and, given enough time, coming as close as desired (but not entirely coinciding) to any given initial condition ([1]).

This phenomenon is signaling certainly that the famous recurrence theorem of Poincar is true for infinite dimensional continuum mechanical systems like that one represented by a bounded domain when subject to non-linear vibrations.

A fundamental question appears in the context of this recurrence phenomenon and concerned to the existence of the time average for the associated non-linear wave motion and naturally leading to the concept of a infinite - dimensional invariant measure for the non-linear wave equation ([1]), the mathematical phenomenon subjacente to the Poincar recurrence theorem.

In this paper, we intend to give mathematical methods rigorous proofs for the validity of the famous Ergodic theorem in a special class of polinomial non-linear and Lipschitz wave motions. Our approach is fully based on Hilbert Space methods previously used to study Dynamical System's in Classical Mechanics which by its turn simplifies enormously the task of constructing explicitly the associated Birkoff–Von Neumann invariant measure on the infinite - dimensional space of wave's motion initial conditions. This study is presented on the main section of this paper, namely section 4 and for the Klein Gordon wave motion.

In section 2, we give a very detailed proof of the RAGE'S theorem, a basic functional analysis rigorous method used in our proposed Hilbert Space generalization of the usual finite - dimensional Ergodic theorem and fully presented in section 3.

In section 5, we complement our studies by commenting the important case of Klein Gordon wave - diffusion under random stirring.

9.2 On the detailed mathematical proof of the R.A.G.E. theorem[1]

In this purely mathematical first section 2 of our study, we intend to present a detailed mathematical proof of the R.A.G.E theorem ([2]) used on the analytical proof of ours of the Ergodic theorem on section 3 for Hamiltonian systems of N-particles.

Let us, thus, start our analysis by considering a self - adjoint operator $\mathcal{L}$ on a Hilbert Space $(H, (,))$ where $H_c(\mathcal{L})$ denotes the associated continuity sub-space obtained from the spectral theorem applied to $\mathcal{L}$. We have the following result:

Proposition (RAGE theorem) 1. *Let* $\psi \in H_c(\mathcal{L})$ *and* $\tilde{\psi} \in H$. *We have the Ergodic limit*

$$\lim_{T\to\infty} \frac{1}{T} \int_0^T \left|(\tilde{\psi}, \exp(i\,t\mathcal{L})\psi)\right|^2 dt = 0 \tag{9.1}$$

Proof: In order to show the validity of the above ergodic limit, let us re-write eq. (9.1) in terms of the spectral resolution of $\mathcal{L}$, namely.

$$I = \frac{1}{T} \int_0^T \left|(\tilde{\psi}, \exp(i\,t\mathcal{L})\psi)\right|^2 dt = \frac{1}{T} \int_0^T \left\{ \left[\int_{-\infty}^{+\infty} e^{+it\lambda} d_\lambda(\tilde{\psi}, E_j(\lambda)\psi)\right] \right.$$
$$\left. \times \left[\int_{-\infty}^{+\infty} e^{-it\mu}\, d_\mu(\tilde{\psi}, E_j(\mu)\psi)\right] \right. \tag{9.2}$$

[1] See Appendix C for a rigorous mathematical proof of the Ergodic Theorem for Wide-Sense Stationary Stochastic Process.

Here we have used the usual spectral representation of $\mathcal{L}$

$$(g, \mathcal{L}\, h) = \int_{-\infty}^{+\infty} \lambda\, d_\lambda\, (g,\, E_j(\lambda)h) \tag{9.3}$$

with $g \in H$ and $h \in$ Dom $(\mathcal{L})$.

Let us remark that the function $\exp(i(\lambda - \mu)t)$ is majorized by the function 1 which, by its turn, is an integrable function on the domain $[0, T] \times R \times R$ with the product measure

$$\frac{dt}{T} \otimes d_\lambda(\tilde{\psi}, E_j(\lambda)\psi) \otimes d_\mu(\tilde{\psi}, E_j(\mu)\psi) \tag{9.4}$$

since

$$\int_0^T \frac{dt}{T} \otimes d_\lambda(\tilde{\psi}, E_j(\lambda)\psi) \otimes d_\mu(\tilde{\psi}, E_j(\mu)\psi) = (\tilde{\psi}, \psi)^2 < \infty \tag{9.5}$$

At this point, we can safely apply the Fubbini theorem for interchange the order of integration in relation to the t-variable which leads to the partial result below

$$I = \int_{-\infty}^{+\infty} \int_{-\infty}^{+\infty} \left(\frac{e^{i(\lambda-\mu)T} - 1}{i(\lambda - \mu)T} \right) d_\lambda \langle \tilde{\psi}, E_j(\lambda)\psi \rangle \otimes d_\mu \langle \tilde{\psi}, E_j(\mu)\psi \rangle \tag{9.6}$$

Let us consider two cases. Firstly we take $\tilde{\psi} = \psi$. In this case, we have the bound

$$\left| \frac{e^{i(\lambda-\mu)T} - 1}{i(\lambda - \mu)T} \right| \leq \left| \frac{2\,\mathrm{sen}(\frac{(\lambda-\mu)}{2}T)}{(\lambda - \mu)T} \right| \leq 1 \tag{9.7}$$

If we restrict ourselvers to the case of $\lambda \neq \mu$, a direct application of the Lebesgue convergence theorem on the limit $T \to \infty$ yields that $I = 0$.

On the other hand, in the case of $\lambda = \mu$ we intend now to show that the set $\mu = \lambda$ is a set of zero measure in R^2 in respect with the measure $d_\lambda \langle \psi, E_j(\lambda)\psi \rangle \otimes d_\mu \langle \psi, E_j(\mu)\psi \rangle$. This can be seen by considering the real function on R

$$f(\lambda) = (\psi, E_j(\lambda)\psi) \tag{9.8}$$

We observe that this function is continuous and non-decreasing since $\psi \in H_c(\mathcal{L})$ and range $E_j(\lambda_1)$**C** range $E_j(\lambda_2)$ for $\lambda_1 \leq \lambda_2$. Now, $f(\lambda)$ is an uniform continuous function on the whole real line R, since for a given $\varepsilon > 0$, there is a constant $\tilde{\lambda}$ such that

$$(\psi, E_j(-\infty, -\tilde{\lambda})\psi) < \frac{\varepsilon}{2} \tag{9.9}$$

$$\|\psi\|^2 - (\psi, E_j(-\infty, \tilde{\lambda})\psi) < \frac{\varepsilon}{2} \tag{9.10}$$

Additionally, $f(\lambda)$ is uniform continuous on the closed interval $[-\tilde{\lambda}, \tilde{\lambda}]$ and $f(\lambda)$ can not make variations greater than $\frac{\varepsilon}{2}$ on $(-\infty, -\tilde{\lambda}]$ and $[+\tilde{\lambda}, \infty)$, besides of being a monotonic function on R. These arguments show the uniform continuity of $f(\lambda)$ on whole line R. Hence we have that for a given $\varepsilon > 0$, exists a $\delta > 0$ such that

$$\Big|(\psi, E_j(\lambda')\psi) - (\psi, E_j(\lambda'')\psi)\Big| \leq \frac{\varepsilon}{\|\psi\|^2} \tag{9.11}$$

for

$$\Big|\lambda' - \lambda''\Big| \leq \delta \tag{9.12}$$

In particular for $\lambda' = \lambda + \delta$ e $\lambda'' = \lambda - \delta$

$$(\psi, E_j(\lambda+\delta)\psi) - (\psi, E_j(\lambda-\delta)\psi) \leq \frac{\varepsilon}{\|\psi\|^2} \tag{9.13}$$

As a consequence, we have the estimate

$$\begin{aligned}
\int_{-\infty}^{+\infty} d_\lambda(E_j(\lambda)\psi, \psi) \int_{\lambda-\delta}^{\lambda+\delta} d_\mu(\psi, E_j(\mu)\psi) \\
&\leq \|\psi\|^2 \times ((\psi, E_j(\lambda+\delta)\psi) - (\psi E_j(\lambda-\delta)\psi)) \\
&\leq \|\psi\|^2 \times \frac{\varepsilon}{\|\psi\|^2} \leq \varepsilon
\end{aligned} \tag{9.14}$$

Let us note that for each $\varepsilon > 0$, there is a set $D_\delta = \{(\lambda, \mu) \in R^2; |\lambda - \mu| < \delta\}$ wich contains the line $\lambda = \mu$ and has measure less than ε in relation to the measure $d_\lambda(E_j(\lambda)\psi, \psi) \otimes d_\mu(E_j(\mu)\psi, \psi)$ as a result of eq. (9.14). This shows our claim that $I = 0$ in our special case.

In the general case of $\tilde{\psi} \neq \psi$, we remark that solely the orthogonal component on the continuity sub-space $H_c(\mathcal{L})$ has a non - vanishing inner product with $\exp(it\mathcal{L})\psi$. By using now the polarization formulae, we reduce this case to the first analyzed result of $I = 0$.

At this point we arrive at the complete R.A.G.E theorem ([2])

Theorem 2. *Let K be a compact operator on $(H, (\ ,\ ))$. We have thus the validity of the Ergodic limit*

$$\lim_{T\to\infty} \frac{1}{T} \int_0^T \|K \exp(i\,t\mathcal{L})\psi\|_H^2 \, dt = 0 \tag{9.15}$$

with $\psi \in H_c(\mathcal{L})$.

We leave the details of the proof of this result for the reader, since any compact operator is the norm operator limit of finite - dimension operators and one only needs to show that

$$\lim_{T\to\infty}\frac{1}{T}\int_0^T \|\sum_{n=0}^{N} c_n(e^{it\mathcal{L}}\psi, e_n)g_n\|_H \, dt = 0 \tag{9.16}$$

for c_n constants and $\{e_n\}, \{g_n\}$ a finite set of vector of $(H, (,))$.

9.3 On the Boltzman Ergodic Theorem in Classical Mechanics as a result of the R.A.G.E theorem

One of the most important statement in Physics is the famous zeroth law of thermodynamics: "any system approaches an equilibrium state". In the classical mechanics frameworks, one begins with the formal elements of the theory. Namely, the phase-space R^{6N} associated to a system of N-classical particles and the set of Hamilton equations

$$\dot{p}_i = -\frac{\partial H}{\partial q_i}; \quad \dot{q}_i = \frac{\partial H}{\partial p_i} \tag{9.17}$$

where $H(q,p)$ is the energy function.

The above cited thermodynamical equilibrium principle becomes the mathematical statement that for each compact support continuous functions $C_c(R^{6N})$, the famous ergodic limit should holds true ([3]).

$$\int_{R^{6N}} d^{3N}q(0)d^{3N}p(0)\left\{\lim_{T\to\infty}\frac{1}{T}\int_0^T f(q(t);p(t))dt\right\} = \eta(f) \tag{9.18}$$

where $\eta(f)$ is a linear functional on $C_c(R^{6N})$ given exactly by the Boltzman statistical weight and $\{q(t),p(t)\}$ denotes the (global) solution of the Hamilton equations (9.12).

For instance:

Theorem 3. *Let $V(q^i)$ be a C^2 function on $\mathbb{R}^{DN}$ with $||\nabla_{q_j}V(q^i)||_{\mathbb{R}^{DN}} \le C(||q||^2_{\mathbb{R}^{DN}} + a^2)^{1/2}$ for suitable positive non-vanishing constants C and a.*

Let $H(p^i,q^i) = \sum_{i=1}^{3N} A_{ij}p_ip_j + V(q^i)$ on $\mathbb{R}^{2DN}$, with A denoting a strictly positive definite matrix. Then for any $\langle p_0, q_0\rangle \in R^{2DN}$, there is a unique C^1 function from $\mathbb{R}$ to $\mathbb{R}^{2DN}$, denoted by $w(t;p_0,q_0)$ and satisfying globally Eq. (9.17) with initial conditions $\langle p_0, q_0\rangle$.

Moreover $w(t;p_0,q_0)$ is an C^1 application of R^{2DN+1} to R^{2DN} (see ref. [1]).

Note that if $V(q^i)$ is a Lipschitzian function on $C^2(R^{DN})$ the theorem necessarily holds true.

We aim at this section point out a simple new mathematical argument of the fundamental eq. (9.18) by means of Hilbert Space Techniques and the R.A.G.E'S theorem. Let us begin by introducing for each initial condition $(q(0),p(0))$, a function $\omega_{q_0,p_0}(t)\equiv(q(t),p(t))$, here $\langle q(t),p(t)\rangle$ is the assumed global unique solution of eq. (9.17) with prescribed initials conditions. Let $U_t: L^2(R^{6N})\to L^2(R^{6N})$ be the unitary operator defined by

$$(U_t f)(q,p)=f(\omega_{q_0 p_0}(t)) \tag{9.19}$$

We have the following theorem (the Liouville's theorem) ([1]).

Theorem 4. *U_t is a unitary one-parameters group whose infinitesimal generator is $-i\bar{L}$, where $-iL$ is the essential self-adjoint operator acting on $C_0^\infty(R^{6N})$ defined by the Poisson bracket.*

$$(Lf)(p,q)=\{f,H\}(q,p)=\sum_{i=1}^{3N}\left(\frac{\partial f}{\partial q_i}\frac{\partial H}{\partial p_i}-\frac{\partial f}{\partial p_i}\frac{\partial H}{\partial q_i}\right)(q,p) \tag{9.20}$$

The basic result we are using to show the validity of the ergodic limit eq. (9.18) is the famous R.A.G.E's theorem exposed on section 1.

Theorem 5. *Let $\phi\in H_c(-i\bar{L})(L^2(R^{6N}))$, here $H_c(-i\bar{L})$ is the continuity sub-space associated to self-adjoint operator $-i\bar{L}$. For every vector $\beta\in L^2(R^{6N})$, we have the result*

$$\lim_{T\to\infty}\frac{1}{T}\int_0^T\left|\langle\beta,U_t\psi\rangle\right|^2 dt=0, \tag{9.21}$$

or equivalently for every $\psi\in L^2(R^{6N})$ and $\beta\in L^2(R^{6N})$

$$\lim_{T\to\infty}\frac{1}{T}\int_0^T\langle\beta,U_t\psi\rangle dt=\langle\beta,\mathcal{P}_{ker(-i\bar{L})}\psi\rangle \tag{9.22}$$

where $\mathcal{P}_{ker(-i\bar{L})}$ is the projection operator on the (closed) sub-space $ker(-i\bar{L})$.

That eq. (9.22) is equivalent to eq. (9.21), is a simple consequence of the Schwartz inequality below written

$$\left|\int_0^T(\langle\beta,U_t\psi\rangle)dt\right|\le\left(\int_0^T(\langle\beta,U_t\psi\rangle)^2dt\right)^{\frac{1}{2}}\left(\int_0^T 1dt\right)^{\frac{1}{2}} \tag{9.23}$$

or

$$\left|\frac{1}{T}\int_0^T \langle\beta, U_t\psi\rangle dt\right| \leq \left(\frac{1}{T}\int_0^T (\langle\beta, U_t\psi\rangle)^2 dt\right)^{\frac{1}{2}} \tag{9.24}$$

As a consequence of eq. (9.23), we can see that the linear functional $\eta(f)$ of the Ergodic theorem is exactly given by (just consider $\beta(p,q) \equiv 1$ on supp of $\mathbb{P}_{\ker(-i\bar{L})}(\psi)$, for $\psi \in C_c(R^{6N})$ arbitrary)

$$\eta(f) = \int_{R^{6N}} dqdp\, \mathbb{P}_{\ker\,(-i\bar{L})}(f)(q,p). \tag{9.25}$$

By the Riesz' s theorem applied to $\eta(f)$, we can re-write (represent) eq. (9.25) by means of a (kernel)-function $h_{\eta(H)}(q,p)$, namely

$$\eta(f) = \int_{R^{6N}} d^{3N}qd^{3N}p \cdot f(p,q) \cdot h_{\eta(H)}(p,q) \tag{9.26}$$

where the function $h_{\eta(H)}(q,p)$ satisfies the relationship

$$\{h_{\eta(H)}, L\} = \sum_{i=1}^{3N}\left(\frac{\partial h_\eta}{\partial q_i}\frac{\partial H}{\partial p_i} - \frac{\partial H}{\partial q_i}\frac{\partial h_\eta}{\partial p_i}\right) = 0 \tag{9.27}$$

or equivalently $h_{\eta(H)}(q,p)$ is a "smooth" function of the Hamiltonian function $H(q,p)$, by imposing the additive Boltzman behavior for $h_{\eta(H)}(q,p)$ namely, $h_{\eta(H_1+H_2)} = h_{\eta(H_1)} \cdot h_{\eta(H_2)}$, one obtains the famous Boltzman weight as the (unique) mathematical output associated to the Ergodic Theorem on Classical Statistical Mechanics in the presence of a thermal reservoir ([4], [5]).

$$h_{\eta(H)}(q,p) = \exp\{-\beta H(q,p)\}\Big/\int d^{3N}qd^{3N}p\,\exp\{-\beta H(q,p)\} \tag{9.28}$$

with β a (positive) constant which is identified with the inverse macroscopic temperature of the combined system after evaluating the system internal energy in the equilibrium state. Note that $\|h_{\eta(H)}\|_{L^2} = 1$ since $\|\eta(f)\| = 1$.

A last remark should be made related to eq. (9.28). In order to obtain this result one should consider the non-zero value in ergodic limit

$$\lim_{T\to\infty}\frac{1}{T}\int_0^T dt(U_t\psi)(q,p)h_{\eta(H)}(q,p) = \langle h_{\eta(H)}, \mathcal{P}_{\ker\,(-i\bar{L})}(\psi)\rangle \tag{9.29}$$

or by a pointwise argument (for every t)

$$(U_{-t}h_{\eta(H)}) \in \mathcal{P}_{\ker\,(-i\bar{L})}, \tag{9.30}$$

that is

$$h_{\eta(H)} \in \mathcal{P}_{\ker\,(-i\bar{L})} \Leftrightarrow Lh = 0. \tag{9.31}$$

9.4 On the invariant ergodic functional measure for non-linear Klein-Gondon wave equations with kinetic trace class operators

Let us start by considering the following initial value non-linear wave equation with a positive trace class operator as a kinetic operator on $\mathbb{R}^D$ ([4]), with $U(t,x) \in C(R^+, L^2(R^N))$ (see Theorem 3 for the discreticized case)[2]

$$U_{tt}(x,t) = -\left\{\left[X_\Omega((-\Delta)^\alpha + m^2)^{-1}X_\Omega\right]^{-1} U\right\}(x,t) + \frac{\delta V(U)}{\delta U}(x,t) \tag{9.32-a}$$

$$\begin{aligned} U(x,0) &= f(x) \in S(\mathbb{R}^D) \\ U_t(x,0) &= g(x) \in S(\mathbb{R}^D) \end{aligned} \tag{9.32-b}$$

Here the Klein Gordon kinetic operator is given by an important set of inverse of class of trace operator on Mathematical Physics of Constructive Field Theory of rigorous path integrals ([4]), and defined by continuum powers of the (strictly positive) Laplacean operator on R^D (or any other strongly uniform elliptic operator) but projected on a given compact domain $\Omega \subset R^D$ through its charactheristic function $I_\Omega(x) \equiv \chi_\Omega(x) \equiv \begin{cases} 1 & x \in \Omega \\ 0 & x \in \Omega^0 \end{cases}$. The integral kernel of such class is given (defined) by

$$T_{\Omega,\alpha,m^2)} = \chi_\Omega(x)((-\Delta)^\alpha + m^2)^{-1}(x,y)\chi_\Omega(y). \tag{9.33}$$

The non negative non linear term $V(U(x,t))$ on our proposed Klein Gordon Model
eqs. (9.32-a)–(9.32-b) is such that it allows to global solutions of the associated initial value Klein Gordon problem. Namely

$$||\nabla_{\mathbb{R}^D} V(q^i)||_{\mathbb{R}^D} \leq C(||q^i||^2_{\mathbb{R}^D} + a^2)^{1/2}. \tag{9.34}$$

We now consider the discreticized (N particle) wave motion Hamiltonian associated to the non-linear Klein-Gordon wave equation (for $\alpha > \frac{D}{2}$)

$$H(p_i, q_i) = \sum_{i=1}^{N} \left\{\frac{p_i^2}{2} + \lambda_i^2 q_i^2 + \widetilde{V}_N(q_1^1, \ldots, q_N^1)\right\}. \tag{9.35}$$

Here

$$U(x,t) = \sum_{i=1}^{\infty} q_i(t)\phi_i(x) \tag{9.36-a}$$

[2]We note that the existence, uniqueness and globality on $t \in [0,\infty]$ of the $U(x,t) \in C([0,T], L^2(R^D))$ through the application of the fixed point theorem ([2]).

$$\left(T_{(\Omega,\alpha,m^2)}\phi_i\right)(x) = \left(\frac{1}{\lambda_i}\right)\phi_i(x) \tag{9.36-b}$$

$$\widetilde{V}_N(q_1,\ldots,q_N) = \int_{\mathbb{R}^D} V\left(\sum_{i=1}^{N} q_i(t)\phi_i(x)\right) d^D x \tag{9.36-c}$$

Note the pointwise convergence for each fixed $t \in R^+$

$$\lim_{N\to\infty}\left(\widetilde{V}_N(q_1,\ldots,q_N)\right) = \int_{R^D} V(U(x,t))d^D x. \tag{9.37}$$

If $F \in C_c(R)$, it holds true also

$$\lim_{N\to\infty}\left\{\int_{R^D}\left[F\left(\sum_{i=1}^{N} q_i(t)\phi_i(x)\right)\right] d^D x\right\}$$

$$\stackrel{\text{(definition)}}{\equiv} \widetilde{F}_N(q_1(t),\ldots,q_N(t)) \xrightarrow{\text{pointwise convergence for each fixed } t} \int_{R^D} F(U(x,t))d^D x \tag{9.38}$$

After these remarks we have our principal result

Theorem 6. *Under the above stateg conditions, we have the Von-Neumann Birkoff result in the infinite dimension case*

$$\lim_{T\to\infty}\left\{\frac{1}{T}\int_0^T\left[\int_{\mathbb{R}^D} F(U(x,t))d^D x\right] dt\right\}$$

$$= \int_{\mathbb{R}^D} d^D x\left\{\int (d_{T_{(\Omega,\alpha,m^2)}}\mu(\varphi(x)))\left(e^{-\beta\int_{\mathbb{R}^D} V(\varphi(x))d^D x}\right) F(\varphi(x))\right\} \tag{9.39}$$

Proof: Let us apply the already proved ergodic theorem for our N-particle system with Hamiltonian function eq. (9.35). Namely

$$\lim_{T\to\infty}\left\{\frac{1}{T}\int_0^T \widetilde{F}_N(q_1(t),\ldots,q_N(t))dt\right\}$$

$$= \frac{1}{Z}\left\{\int_{\mathbb{R}^{ND}} \widetilde{F}_N(q_1,\ldots,q_N)\exp\left[-\beta\left(\sum_{i=1}^{N}\lambda_i^2 q_i^2 + \widetilde{V}_N(q_1,\ldots,q_N)\right)\right]\right\} \tag{9.40}$$

where Z is the invariant ergodic measure normalization factor (given explicitly by considering the integrand on the right-hand side of eq. (9.40) with $\widetilde{F}_N(q_1,\dots,q_N)\equiv 1$).

At this point we can take safely the limit of $N\to\infty$ on the cylindrical measures of the right-hand side of eq. (9.40). Since the operator T_Ω is a class trace operator on $L^2(\mathbb{R}^N)$ (for $\alpha>\frac{N}{2}$, and thus by Minlos theorem defining a σ-measure on $L^2(R^N)$ ([4]).

As a consequence we have the infinite dimensional result below written

$$\lim_{N\to\infty}\left\{\lim_{T\to\infty}\left[\frac{1}{T}\int_0^T \widetilde{F}_N(q_1(t),\dots,q_N(t))dt\right]\right\}$$

$$=\int_{R^D}\left\{\int_{L^2(\mathbb{R}^D)} d_{T_{(\Omega,\alpha,m^2)}}\mu(\varphi(x))\cdot e^{-\beta\int_\Omega V(\varphi(x))d^Dx}F(\phi(x))\right\}. \quad (9.40\text{-a})$$

It is straightforward to argument that the left-hand side rigorous mathematical object is the physicist ergodic limit for the infinite particle limit of our Hamiltonian system and taking as a definition for us

$$\lim_{N\to\infty}\left\{\lim_{T\to\infty}\left[\frac{1}{T}\int_0^T \widetilde{F}_N(q_1(t),\dots,q_N(t))dt\right]\right\}$$

$$\overset{\text{(formally)}}{=}\lim_{T\to\infty}\left[\frac{1}{T}\int_0^T\left\{\lim_{N\to\infty}\widetilde{F}_N(q_1(t),\dots,q_N(t))\right\}dt\right]$$

$$\overset{\text{(physicist's definition)}}{\equiv}\lim_{T\to\infty}\left[\frac{1}{T}\int_0^T\left(\int_{\mathbb{R}^D}F(U(x,t))d^Dx\right)dt\right]. \quad (9.41)$$

Exactly mathematical conditions for interchange the above (N,T) limits still remains an open problem in the subject ([5]).

In the important case of $N=1$ and $\alpha=1$ and exponentially cut-off polinomial Klein-Gordon equation on $C^2([0,\ell]\times R^+)$ with vanishing mass $m^2=0$, for $0\le x\le\ell$ (the usual wave equation)

$$\frac{\partial^2 U(x,t)}{\partial^2 t}=\frac{d^2}{dx}(U(x,t))+\left(\frac{\delta V(U)}{\delta U}\right)(x,t)$$

$$U(0,\ell)=U(0,t)=0$$

$$U(x,0)=f(x)\in C^2([0,\ell]);f^{(3)}(x)$$
$$\in L^2([0,\ell]);f(\ell)=f(0);f^{(2)}(0)=f^{(2)}(\ell)=0$$

$$U_t(x,0) = g(x) \in C^1([0,\ell]); g^{(2)}(x) \in L^2([0,\ell]); g(0) = g(\ell) = 0. \quad (9.42)$$

Here the exponentially cut-off interaction (a Lipschitzian function) is explicitly given by ($\varepsilon \in \mathbb{R}^1$)

$$V * U(x,t)) = \sum_{j=1}^{M} ((\lambda_j^{2j}) \exp(-\varepsilon(U(x,t))^2))(U(x,t))^{2j}), \quad (9.43)$$

one can substitute the $T_{(\Omega,\alpha,m^2)}$ operator on the two-dimensional Klein-Gordon wave equation by the usual one-dimensional operator $-\frac{d^2}{dx^2}$, but now defined on $H^2(\Omega)$.

It leads thus to the important result that the associated Von-Neumann-Birkoff invariant measure is given by the usual Wiener-Kac path measure defined for closed Wiener trajectories with the marked fixed point $X(0) = X(1) = x$

$$\lim_{T\to\infty} \left[\frac{1}{T} \left(\int_{\mathbb{R}} F(U(x,t))dx \right) dt \right]$$

$$\int_0^\ell dx \Big\{ \int_{\overbrace{\{C([0,1]), X(0)=X(1)=x\}}^{\text{Wiener closed trajectories}}} d^{\text{Wiener}}\mu[X(\sigma)]$$

$$\times \exp\left\{ -\beta \int_0^1 V(X(\sigma))d\sigma \right\} F(X(\sigma)) \Big\} \quad (9.44)$$

Note that one can consider the case $m = 0$, since the operator $\left(-\frac{d^2}{dx^2}\right)^{-1}$ with Dirichlet Boundary conditions is a trace class positive operator on $L^2(\Omega)$, for Ω any closed interval on R^1.

Concerning the general case of a strongly non-negative elliptic operator A of order $2m$ associated to the free vibration of a domain Ω with compact closure in a general space $\mathbb{R}^D$ (with the condition U) ([6]) we still have the same results above exposed

$$U_{tt} = -\left\{ \left(I_\Omega[(A)^{+\alpha} + m^2]^{-1} I_\Omega\right)^{-1} U \right\}(x,t) + \frac{\delta V(U)}{\delta U}(x,t)$$

$$U(x,0) = f(x) \in C^2(\Omega) \cap C^1(\overline{\Omega})$$

$$V_t(x,0) = g(x) \in C^3(\Omega) \cap C^1(\overline{\Omega}). \quad (9.45)$$

Here A should satisfy the Garding coerciviness condition and given explicitly by

$$A = \sum_{\substack{|p|\le m \\ |q|\le m}} (-1)^{|p|} D_x^p (\mathcal{A}_{pq}(x) D_x^q \quad (9.46)$$

the Garding condition reads as of as:

$$\text{Real}(Au, u)_{L^2(\Omega)} \geq C_0(\Omega)||U||_{H^{2m}(\Omega)}. \tag{9.47}$$

At this point it appears important to point out that even if the limit of infinite volume $\chi_\Omega(x) \to 1$ is immediate on the Klein-Gordon wave equation eq. (9.32-a), quiet opposite to the case of the kinetic operator being $(-\Delta_\Omega)^\alpha(1 - \mathbb{P}_{\ker(\Delta_\Omega)})$; defined by the usual espectral power of the self-adjoint Laplacean operator projected out of the zero modes (the vector space of the harmonic functions on Ω); this infinite volume limit on the associated Birkoff-Von Neumann ergodic invariant measure is highly non trivial since its support in this case of infinite volume (if any) would be fully the L. Schwartz distribution space ([4], [7], [8], [9]) and Appendix A of this paper).

The same remarks apply to the $\alpha \to 1$ limit.

It is thus conjectured by us that the $\alpha \to 1$ and $\chi_\Omega(x) \to 1$ limits shouldbe taken only in the ergodic averages as it is usually expected on Constructive field theory ([9], [10]) (see the right-hand side of eq. (9.39), Theorem 6) and solely after considering the infinite renormalization problem issues ([9], [10], [11]).

9.5 An Ergodic theorem in Banach Spaces and Applications to Stochastic-Langevin Dynamical Systems

In this complementary section 5, we intend to present our approach to study long-time ergodic behavior of infinite-dimensional dynamical systems by analyzing the somewhat formal diffusion equation with polinomial terms and driven by a white noise stirring ([5]).

In order to implement such studies, let us present the author's generalization of the R.A.G.E theorem for a contraction self-adjoint semi-group $T(t)$ on a Banach space X. We have, thus, the following theorem.

Theorem 7. *Let $f \in X$. We have the ergodic generalized theorem*

$$\lim_{T\to\infty} \frac{1}{T} \int_0^T dt(e^{-tA}f) = \mathcal{P}_{ker(A)}(f) \tag{9.48}$$

where A is the infinitesimal generator of $T(t)$.

Proof: As a first step, one should consider eq. (9.48) re-written in terms of the "resolvent operator of A" by means of a Laplace Transform (The

Hile-Yosida - Dunford Spectral Calculus)

$$\lim_{T\to\infty} \frac{1}{T} \int_0^T dt \left\{ \frac{1}{2\pi i} \int_{-i\infty}^{i\infty} dz e^{zt}((z+A)^{-1}f) \right\}. \tag{9.49}$$

Now it is straightforward to apply the Fubbini theorem to exchange the order of integrations (dt, dz) in eq. (9.49) and get, thus, the result

$$\lim_{T\to\infty} \frac{1}{T} \int_0^T dt(e^{-tA}f) = \lim_{z\to 0^-} +((z+A)^{-1}f). \tag{9.50}$$

The $z \to 0^-$ limit of the integral of eq. (9.50) (since Real $(z) \subset \rho(A) \subset (-\infty, 0)$) can be evaluated by means of saddle point techniques applied to Laplace's transforms. We have the following result

$$\begin{aligned} &\lim_{z\to 0^-} ((z+A)^{-1}f) \\ &= \lim_{z\to 0^-} \int_0^\infty e^{-zt}(e^{-tA}f)\, dt \\ &= \lim_{t\to\infty} (e^{-tA}f) = \mathcal{P}_{ker(A)}(f). \end{aligned} \tag{9.51}$$

Let us apply the above theorem to the Lengevin Equation. Let us consider the Fokker-Planck equation associated to the following Langevin Equation

$$\frac{dq^i}{dt}(t) = -\frac{\partial V}{\partial q_i}(q^j(t)) + \eta^i(t) \tag{9.52}$$

where $\{\eta^i(t)\}$ denotes a white-noise stochastic time process representing the thermal coupling of our mechanical system with a thermal reservoir at temperature T. Its two point function is given by the "Fluctuation-Dissipation" theorem

$$\langle \eta^i(t)\eta^j(t')\rangle = kT\delta(t-t'). \tag{9.53}$$

The associated Fokker-Planck equation associated do eq. (9.52) has the following explicitly form

$$\frac{\partial P}{\partial t}(q^i, \bar{q}^i; t) = +(kT)\Delta_{q_i} P(q^i, \bar{q}^i; t) + \nabla_{q_i}(\nabla_{q_i} V \cdot P(q^i, \bar{q}^i; t)) \tag{9.54}$$

$$\lim_{t\to 0+} P(q^i, \bar{q}^i; t) = \delta^{(N)}(q^i - \bar{q}^i). \tag{9.55}$$

By noting that we can associate a contractive semi-group to the initial value problem eq. (9.55) in the Banach Space $L^1(R^{3N})$, namely:

$$\int P(q^i, \bar{q}^i, t) f(\bar{q}^i) d\bar{q}^i = (e^{-tA}f). \tag{9.56}$$

Here the closed positive accretive operator A is given explicitly by

$$-A(\cdot) = kT\Delta_{q_i}(\cdot) + \nabla_{q_i}[(\nabla_{q_i}V)(\cdot)] \tag{9.57}$$

and acts firstly on $X = C_\infty(R^{3N})$. It is instructive to point out that the perturbation accretive operator $B = \nabla_{qi} \cdot [(\nabla_{qi}V) \cdot (\cdot)]$ on $C_\infty(R^{3N})$ with $V(q^i) \in C_c^\infty(R^{3N})$ is such that it satisfies the estimate on $S(R^{3N})$: $\|Bf\|_{S(R^{3N})} \leq a\|\Delta f\|_{S(R^{3N})} + b\|f\|_{S(R^{3N})}$ with $a > 1$ and for some b. As a consequence $A(\cdot) = kT\Delta_{qi}(\cdot) + (B(\cdot))$ generates a contractive semi-group on $C_\infty(R^{3N})$ or by an extension argument on the whole $L^1(R^{3N})$ since the L^1-closure of $C_\infty(R^{3N})$ is the Banach space $L^1(R^{3N})$.

At this point we may apply our Theorem 7 to obtain the Langevin-Brownian Ergodic theorem applied to our Fokker-Planck equation

$$\begin{aligned}\lim_{T\to\infty} \frac{1}{T}\int_0^T dt \left(\int_{R^{3N}} dq(e^{-tA}f)(q)\right) &= \int_{R^{3N}} dq \mathbb{P}_{ker(A)}(f)(q) \\ &= \int_{R^N} d^N\bar{q}^i\, f(\bar{q}^i)\mathcal{P}^{eq}(\bar{q}^i)\end{aligned} \tag{9.58}$$

where the equilibrium probability distribution is given explicitly by the **unique normalizable** element of the closed sub-space ker(A).

$$O = +(kT)\Delta_{q_i}\mathcal{P}^{eq}(\bar{q}^i) + \nabla_{\bar{q}^i}[(\nabla_{\bar{q}^i}V)\mathcal{P}(\bar{q}^i)] \tag{9.59}$$

or exactly, we have the Boltzman's weight for our equilibrium Langevin-Brownian probability distribution[3]

$$\mathcal{P}^{eq}(\bar{q}^i) = \exp\left\{-\frac{1}{kT}(V(\bar{q}^i))\right\}. \tag{9.60}$$

For the general Langevin equation in the complete phase space $\{q_i, p_i\}$ as in the bulk of this note, one should re-obtains the complete Boltzman statistical weight as the equilibrium ergodic probability distribution.

$$\mathcal{P}^{eq}(\bar{q}^i, \bar{p}^i) = \frac{1}{Z}\exp\left\{-\frac{1}{kT}\left[\sum_{l=1}^{N}\frac{1}{2}(\bar{p}^i)^2 + V(\bar{q}^i)\right]\right\} \tag{9.61}$$

with the normalization factor

$$Z = \int_{R^N} d\bar{p}^i \int_{R^N} d\bar{q}^i\, \mathcal{P}^{eq}(\bar{q}^i, \bar{p}^i) \tag{9.62}$$

Let us apply the above exposed results for the initial-value diffusion equation on $\mathbb{R}^D$ $(\alpha > \frac{D}{2})$

$$U_t(x,t) = -\{[\chi_\Omega((-\Delta)^{+\alpha} + m^2)^{-1}\chi_\Omega]^{-1}U\}(x,t)$$

[3]This result has the conceptual advantage of explaining the Boltzman equilibrium statistic weight directly from the Ergodic Theorem.

$$+\left(\frac{\delta V(U)}{\delta U}\right)(x,t)+\eta(x,t)$$

$$U(x,0)=f(x)\in L^2(\mathbb{R}^D)$$

$$\eta(x,t)=\sum_{i=1}^{\infty}\eta_i(t)\phi_i(x). \tag{9.63}$$

By using the same discretisized eigenfunction expansion of section 4, one obtains the associated invariant ergodic measure associated to the non-linear diffusion equation (9.63)

$$d^{\text{inv}}\mu(\varphi(x))=(d_{T_{(\Omega,\alpha,m^2)}}\mu(\varphi))\times e^{-(\beta V(\varphi(x)))} \tag{9.64}$$

9.6 Appendix A: The existence and uniqueness results for some polinomial wave motions in 2D

In this technical Appendix A, we give an argument for the global existence and uniqueness solution of the Hamiltonian motion equations associated firstly to eq. (32) - section 3 and by secondly to eq. (53) - section 3 at the infinite volume limit and for $\alpha=1$. Related to the two-dimensional case, let us equivalently show the weak existence and uniqueness of the associated continuum non-linear polinomial wave equation in the domain $(-a,a)\times R^+$.

$$\frac{\partial^2 U(x,t)}{\partial^2 t}-\frac{\partial^2 U(x,t)}{\partial^2 x}+g(U(x,t))^{2k-1}=0 \tag{A-1}$$

$$U(-a,t)=U(a,t)=0$$
$$U(x,0)=U_0(x)\in H^1([-a,a]) \tag{A-2}$$

$$U_t(x,0)=U_1(x)\in L^2([-a,a]) \tag{A-3}$$

Let us consider the Galerkin approximants functions to eq. (A-2)–eq. (A-3) as given below

$$\overline{U}_n(t)\equiv\sum_{\ell=1}^{n}U_\ell(t)\ \text{sen}\left(\frac{\ell\pi}{a}x\right) \tag{A-4}$$

Since there is a $\gamma_0(a)$ positive such that

$$\left(-\frac{d^2}{d^2x}\overline{U}_n,\overline{U}_n\right)_{L^2([-a,a])}\geq\gamma_0(a)(\overline{U}_n,\overline{U}_n)_{H^1([-a,a])} \tag{A-5}$$

we have the a priori estimate for any t

$$0 \le \varphi(t) \le \varphi(0) \tag{A-6}$$

with

$$\varphi(t) = \frac{1}{2}\|\dot{\overline{U}}_n(t)\|^2_{L^2} + \gamma_0(a)\|\overline{U}_n\|^2_{H^1} + \frac{1}{2k}\|\overline{U}_n\|^{2k}_{L^2} \tag{A-7}$$

As a consequence of the bound eq. (A-7), we get the bounds for any given T (with A_i constants)

$$\sup_{0\le t\le T} \text{ess}\|\overline{U}_n\|^2_{H^1(-a,a)} \le A_1 \tag{A-8}$$

$$\sup_{0\le t\le T} \text{ess}\|\dot{\overline{U}}_n\|^2_{L^2(-a,a)} \le A_2 \tag{A-9}$$

$$\sup_{0\le t\le T} \text{ess}\|\overline{U}_n\|^{2k}_{L^{2k}(-a,a)} \le A_3 \tag{A-10}$$

By usual functional - analytical theorems on weak-compactness on Banach-Hilbert Spaces, one obtains that there is a sub-sequence $\overline{U}_n(t)$ such that for any finite T

$$\overline{U}_n(t)\overset{*}{\overrightarrow{(weak-star)}}\overline{U}(t) \qquad \text{in} \qquad L^\infty([0,T],H^1(-a,a)) \tag{A-11}$$

$$\dot{\overline{U}}_n(t)\overset{*}{\overrightarrow{(weak-star)}}\overline{v}(t) \qquad \text{in} \qquad L^\infty([0,T],L^2(-a,a)) \tag{A-12}$$

$$\overline{U}_n(t)\overset{*}{\overrightarrow{(weak-star)}}\overline{p}(t) \qquad \text{in} \qquad L^\infty([0,T],L^{2k}(-a,a)) \tag{A-13}$$

At this point we observe that for any $p > 1$ (with $\tilde{A}_i$ constants) and $T < \infty$ we have the relationship below

$$\int_0^T \|\overline{U}_n\|^p_{H^1(-a,a)}dt \le T(A_1)^{\frac{p}{2}} \Leftrightarrow \int_0^T \|\overline{U}_n\|^p_{L^2(-a,a)} \le \tilde{A}_1 \tag{A-14}$$

$$\int_0^T \|\dot{\overline{U}}_n\|^p_{L^2(-a,a)}dt \le T(A_2)^{\frac{p}{2}} \Leftrightarrow \int_0^T \|\dot{\overline{U}}\|^p_{L^{2k}(-a,a)}dt \le \tilde{A}_2 \tag{A-15}$$

since we have the continuous injection below

$$H^1(-a,a) \hookrightarrow L^2(-a,a) \tag{A-16}$$

$$L^{2k}(-a,a) \hookrightarrow L^2(-a,a) \tag{A-17}$$

As a consequence of the Aubin-Lion theorem ([4]), one obtains straightforwardly the strong convergence on $L^P((0,T), L^2(-a,a))$ togheter with the almost everywhere point wise equalite among the solutions candidate

$$\overline{U}_n \longrightarrow \overline{U}(t) = \overline{p}(t) = \int_0^t \overline{v}(s)ds \tag{A-18}$$

By the Holder inequalite applied to the pair (q,k)

$$\|U_n - \overline{U}\|_{L^q(-a,a)} \le \|U_n - \overline{U}\|^{1-\theta}_{L^2(-a,a)} \|U_n - \overline{U}\|^{\theta}_{L^{2k}(-a,a)} \tag{A-19}$$

with $0 \le \theta \le 1$

$$\frac{1}{q} = \frac{1-\theta}{2} + \frac{\theta}{2k} \tag{A-20}$$

in particular with $q = 2k-1$, one obtains the strong convergence of $U_n(t)$ in the general Banach space $L^\infty((0,T), L^{2k-1}(-a,a))$, with $\overline{\theta} = \frac{k}{1-k}\left(\frac{3-2k}{2k-1}\right)$

As a consequence of the above obtained results, one can pass safely the weak limit on

$C^\infty((0,T), L^2(-a,a))$

$$\lim_{n\to\infty}\left\{\frac{d^2}{d^2t}(\overline{U}_n, v)_{L^2(-a,a)} + \left(-\frac{d^2}{d^2x}\overline{U}_n, v\right)_{L^2(-a,a)} + g(\overline{U}_n^{2k-1}, v)_{L^2(-a,a)}\right\}$$

$$= \frac{d^2}{d^2t}(\overline{U}, v)_{L^2(-a,a)} + \left(-\frac{d^2}{d^2x}\overline{U}, v\right)_{L^2(-a,a)} + g(\overline{U}^{2k-1}, v)_{L^2(-a,a)} = 0 \tag{A-21}$$

for any $v \in C^\infty((0,T), L^2(-a,a))$

At this point we sketchy a somewhat rigorous argumont to prove the problem's uniqueness.

Let us consider the hypothesis that the finite function

$$a(x,t) = \frac{((\overline{U})^{2k+1}(x,t) - (\overline{v})^{2k+1}(x,t))}{(\overline{U}(x,t) - \overline{v}(x,t))} \tag{A-22}$$

is essentially bounded on the domain $[0,\infty) \times (-a,a)$ where $\overline{U}(x,t)$ and $\overline{v}(x,t)$ denotes, two hypothesized different solutions for the 2D-polinomial wave equation eq. (A-1)–eq. (A-3). It is straightforward to see that its difference $W(x,t) = (\overline{U} - \overline{v})(x,t)$ satisfies the "Linear" wave equation problem

$$\frac{\partial^2 W}{\partial^2 t} - \frac{\partial^2 W}{\partial x} + (a\, W)(x,t) = 0 \tag{A-23}$$

$$W(0) = 0 \tag{A-24}$$

$$W_t(0) = 0 \tag{A-25}$$

At this point we observe the estimate (where $H^1 \hookrightarrow L^2$!)

$$\begin{aligned}
&\frac{1}{2}\frac{d}{dt}\left(\|\frac{d}{dt}W\|^2_{L^2(-a,a)} + \|W\|^2_{H^1(-a,a)}\right) \\
&\le \|a\|_{L^\infty((0,T)\times(-a,a))} \times \left(\|W\|_{L^2(-a,a)} \times \|\frac{dW}{dt}\|_{L^2)-a,a)}\right) \\
&\le M\left(\|\frac{dW}{dt}\|^2_{L^2(-a,a)} + \|W\|^2_{L^2(-a,a)}\right) \\
&\le M\left(\|\frac{dW}{dt}\|^2_{L^2(-a,a)} + \|W\|^2_{H^1(-a,a)}\right)
\end{aligned} \tag{A-26}$$

which after a application of the Gronwall's inequalite give us that

$$\begin{aligned}
&\left(\|\frac{d}{dt}W\|^2_{L^2(-a,a)} + \|W\|^2_{H^1)-a,a)}\right)(t) \\
&\le \left(\|\frac{dW}{dt}\|^2_{L^2(-a,a)}(0) + \|W\|^2_{H^1(-a,a)}(0)\right) = 0
\end{aligned} \tag{A-27}$$

which proves the problem's uniqueness under the not proved yet hypothesis that in the two-dimensional case (at least for compact support infinite differentiable initial conditions)

$$\sup_{\substack{x\in(-a,a)\\ t\in[0,\infty)}} \|a(x,t)\| \le M. \tag{A-28}$$

It is thus expected (but not proven) that the associated invariant measure would be given by

$$\int_{-a}^{a} d\nu \left\{\int_{D'((0,1))} d_{(-\frac{d^2}{dx^2})^{-1}}\mu[X(\sigma)]e^{-\beta\{g\int_0^1 (X(\sigma))^{2k}d\sigma\}}F(X(\sigma))\right\}. \tag{A-29}$$

Here the formal distributional operator $(-\frac{d^2}{dx^2})^{-1}$ acts on $D((0,1))$ to $D'((0,1))$ through the rule

$$\begin{aligned}
\left(-\frac{d^2}{dx^2}\right)^{-1} : D((0,1)) &\longrightarrow D'_{T_{(-\frac{d^2}{dx^2})^{-1}}}((0,1)) \\
f &\longrightarrow T[f]
\end{aligned} \tag{A-30}$$

Namely

$$T[f](g) = T_{-((\frac{d^2}{dx^2})^{-1}f)}(g) = \int_0^1\int_0^1 \left(f(s)\left(-\frac{d^2}{dx^2}\right)^{-1}(s,t)g(t)\right)dsdt. \tag{A-31}$$

Note that through the Minlos Theorem, one has rigorously ([4]):

$$\exp\left\{-\frac{1}{2}T_{(-\frac{d^2}{dx^2})^{-1}(j)}(j)\right\} = \int_{D'((0,1))} d_{(-\frac{d^2}{dx^2})^{-1}}\mu(X)\exp\left\{i\,X(j)\right\}. \tag{A-32}$$

The above displayed formulae is our mathematical re-wording of our final comments on section 4 of this paper.

9.7 Appendix B: The Ergodic theorem for Quantized wave propagations

In this appendix B we intend to present our results on the quantum statistical case, just for completeness of the exposition of our ideas and proposed mathematical technique.

In quantum theory the equations of motions are operator equations in terms of a given fixed self-adjoint Hamiltonian $\hat{H}_{op}(\hat{p}_i, \hat{q}_i)$

$$\frac{d\hat{q}_i(t)}{dt} = \frac{i}{\hbar}[\hat{H}_{op}, \hat{q}_i](t) \tag{B-1}$$

$$\frac{d\hat{p}_i(t)}{dt} = \frac{i}{\hbar}[\hat{H}_{op}, \hat{p}_i](t). \tag{B-2}$$

Added of initial operatorial conditions

$$\hat{q}_i(0) = \hat{q}_i^{(0)}; \quad \hat{p}_i(0) = \hat{p}_i^{(0)}. \tag{B-3}$$

Proceeding exactly as in section 3 of this paper, the self-adjoint infinitosimal generator of the time-evolution quantum mechanical operator $\hat{L}$ satisfy the quantum mechanical Poisson bracket on the system observable algebra of the trace class operator $\hat{f}(\hat{p}_i, \hat{q}_i)$. Namely

$$L(\hat{f}(\hat{p}_i, \hat{q}_i)) = [\hat{f}(p_i, q_i), \hat{H}_{op}]. \tag{B-4}$$

By imposing again the additive bounded operator Boltzam behavior for the operatorial kernel for the linear functional $\hat{\eta}_f$ acting on trace class operator system observable in the norm topology, one arrives at the N-particle quentum mechanical ergodic theorem for a given quantum mechanical system observable described by a trace class operator $\hat{\mathcal{O}}(\hat{p}^i, \hat{q}^i)$ ([12], [13])

$$\lim_{T\to\infty}\frac{1}{T}\left\{\int_0^T Tr\{\hat{\mathcal{O}}(\hat{p}^i(t), \hat{q}^i(t))\}dt\right\}$$

$$= Tr_{(\mathcal{H}\otimes\mathcal{H}\otimes\ldots\mathcal{H})_n}\left\{e^{-\beta\hat{H}(\hat{p}_i,\hat{q}_i)}\hat{\mathcal{O}}(\hat{p}_i, \hat{q}_i)\right\} Tr_{(\mathcal{H}\otimes\mathcal{H}\otimes\ldots\mathcal{H})_n}\{e^{-\beta\hat{H}(\hat{p}_i,\hat{q}_i)}\}. \tag{B-5}$$

For a second-quantized Klein-Gordon field associated to eqs. (A-32)–(B-3), the wave field has the operatorial expansion

$$\hat{U}(x,t) = \sum_{i=1}^{N} \hat{q}_i(t)\phi_i(x). \tag{B-6}$$

Here $\hat{q}_i(t)$ are operators acting on the Hilbert space of N-free particles $(\mathcal{H}\otimes\mathcal{H}\otimes\ldots\mathcal{H})_N$ and $\phi_i(x)$ are the eigenfunctions of the $T_{(\Omega,\alpha,m^2)}$ operator (see eqs. (A-36)–(B-6)).

Proceeding as in the bulk of this paper, one arrives at the Bosonic Bell functional integral representation ([11]) for the quantum ergodic theorem

$$\lim_{T\to\infty}\left\{\int_0^T Tr\{\hat{\mathcal{O}}(\hat{U}(x,t))\}dt\right\}$$

$$= \lim_{N\to\infty}\left\{\lim_{T\to\infty}\int_0^T Tr\{\hat{\mathcal{O}}(\hat{q}_1(t),\ldots,\hat{q}_N(t))\}dt\right\}$$

$$= \frac{1}{2}\Bigg\{\int_{\substack{L^2([0,\beta]\times\mathbb{R}^D)\\ \phi(x,t+\beta)=\phi(x,t)}} d_{(-\frac{d^2}{dt^2}+T^{-1}_{\Omega,\alpha,m^2})^{-1}}\mu(\phi(x,t))$$

$$\times\exp\left\{-\left[\int_0^\beta\left(\int_{\mathbb{R}^D} V(\phi(x,t))d^Dx\right)dt\right]\right\}\times\mathcal{O}(\phi(x,t))\Bigg\}. \tag{B-7}$$

Here the cilindrical measure $d_{(-\frac{d^2}{dt^2}+T^{-1}_{\Omega,\alpha,m^2})^{-1}}\mu(\phi(x,t))$ is defined by its generating functional

$$Z(j) = \exp\left\{-\frac{1}{2}\left\langle j,\left(-\frac{d^2}{dt^2}+T^{-1}_{\Omega,\alpha,m^2}\right)^{-1} j\right\rangle_{L^2([0,R]\times\mathbb{R}^D)}\right\}$$

$$= \int_{(-\frac{d^2}{dt^2}+T^{-1}_{\Omega,\alpha,m^2})^{-1}} d\mu(\phi(x,t))\exp\left\{i\langle\phi(x,t),j(x,t)\rangle_{L^2([0,R]\times\mathbb{R}^D)}\right\}. \tag{B-8}$$

Note that we have the bound

$$Tr_{L^2([0,\beta]\times\mathbb{R}^D)}\left\{\left(-\frac{d^2}{dt^2}+T^{-1}_{\Omega,\alpha,m^2}\right)^{-1}\right\}$$

$$\leq \sum_{n=u}^{\infty}\left(\left[m^2+\frac{n^2\pi^2}{\beta^2}\right]^{-[2\alpha+\frac{D}{2}]}\times\frac{1}{2}B\left(\frac{D}{2},\alpha-\frac{D}{2}\right)\right)<\infty \tag{B-9}$$

which is convergent for $\alpha > \frac{D}{2}$ as supposed.

A more detailed analysis of the quantum case will appears elsewhere.

9.8 Appendix C: A Rigorous Mathematical proof of the Ergodic theorem for Wide-Sense Stationary Stochastic Process

Let us start our appendix by considering a wide-sense mean continuous stationary real-valued process $\{X(t), -\infty < t < \infty\}$ in a probability space $\{\Omega, d\mu(\lambda), \lambda \in \Omega\}$. Here Ω is the event space and $d\mu(\lambda)$ is the underlying probability measure.

It is well-know that one can always represent the above mentioned wide-sense stationary process by means of a unitary group on the Hilbert Space $\{L^2(\Omega), d\mu(\lambda)\}$. Namely [in the quadratic-mean sense in Engineering jargon]

$$X(t) = U(t)X(0) = \int_{-\infty}^{+\infty} e^{iwt} d(E(w)X(0)) = e^{iHt}(X(0)) \tag{C-1}$$

here we have used the famous spectral Stone-theorem to re-write the associated time-translation unitary group in terms of the spectral process $dE(w)X(0)$, where H denotes the infinitesimal unitary group operator $U(t)$. We have supposed too that the σ-algebra generated by the $X(t)$-process is the whole measure space Ω, and $X(t)$ is a separable process.

Let us, thus, consider the following linear continuous functional on the Hilbert (complete) space $\{L^2(\Omega), d\mu(\lambda)\}$ - the space of the square integrable random variables on Ω

$$L(Y(\lambda)) = \lim_{T\to\infty} \frac{1}{2T} \int_{-T}^{T} dt E\{Y(\lambda)\,\overline{X(t,\lambda)}\,\}. \tag{C-2}$$

By a straightforward application of the R.A.G.E. theorem, namely:

$$\begin{aligned} L(Y(\lambda)) &= \int_{\Omega} d\mu(\lambda) Y(\lambda) \left\{ \lim_{T\to\infty} \frac{1}{2T} \int_{-T}^{T} dt e^{-iwt}\, \overline{dE(w)X_0(\lambda)} \right\} \\ &= \lim_{T\to\infty} \frac{1}{2T} \int_{-T}^{T} E\{Y e^{-iHt}\, \overline{X}\} dt \\ &= \int_{\Omega} d\mu(\lambda) Y(\lambda)\, \overline{dE(0)X(0,\lambda)} \\ &= E\{Y(\lambda)\, \overline{P_{\text{Ker}(H)}(X(0,\lambda))}\}. \end{aligned} \tag{C-3}$$

Here $P_{\text{Ker}(H)}$ is the (ortoghonal projection) ou the kernel of the unitary-group infinitesimal generator H (see eq. (C-1)).

By a straighforward application of the Riesz-representation theorem for linear functionals on Hilbert Spaces, one can see that $\overline{P_{\text{Ker}(H)}(X(0))}\, d\mu(\lambda)$

is the searched time-independent ergodic-invariant measure associated to the ergodic theorem statement, i.e. For any square integrable time independent random variable $Y(\lambda) \in L^2(\Omega, d\mu(\lambda))$, we have the ergodic result $(X(0,\lambda) = X(0)$.

$$\lim_{T\to\infty} \frac{1}{2T}\int_{-T}^{T} dtE\{X(t)\overline{Y}\} = E\{P_{\mathrm{Ker}(H)}(X(0))\overline{Y}\}. \tag{C-4}$$

In general grounds, for any real bounded borelian function it is expected the result (not proved here)

$$\lim_{T\to\infty} \frac{1}{2T}\int_{-T}^{T} dtE\{f(X(t))\overline{Y}\} = E\{P_{\mathrm{Ker}(H)}f(X(0))\cdot\overline{Y}\}. \tag{C-5}$$

For the auto-correlation process function, we still have the result for the translated time ζ fixed (the lag time) as a direct consequence of eq. (C-1) or the process' stationarity property

$$\lim_{T\to\infty} \frac{1}{2T}\int_{-T}^{T} dtE\{X(t)X(t+\zeta)\} = E\{X(0)X(\zeta)\}. \tag{C-6}$$

It is important remark that we still have the probability average inside the ergodic time-averages eqs. (C-4)–(C-6). Let us call the reader attention that in order to have the usual Ergodic like theorem result - without the probability average E on the left-hand side of the formulae, we proceed [as it is usually done in probability text-books] by analyzing the probability convergence of the single sample stochastic-variables below [for instance]

$$\eta_T = \frac{1}{2T}\int_{-T}^{T} f(X(t))dt \tag{C-7}$$

$$R_T(\zeta) = \frac{1}{2T}\int_{-T}^{T} dtX(t)X(t+\zeta). \tag{C-8}$$

It is straightforward to show that if $E\{f(X(t))f(X(t+\zeta))\}$ is a bounded function of the time-lag, or, if the variance below written goes to zero at $T\to\infty$

$$\sigma_T^2 = \lim_{T\to\infty}\frac{1}{4T^2}\int_{-T}^{T} dt_1\int_{-T}^{T} dt_2[E\{X(t_1)X(t_1+\zeta)X(t_2+\zeta))$$

$$-E\{X(t_1)X(t_1+\zeta)\}E\{X(t_2)X(t_2+\zeta)))] = 0, \tag{C-9}$$

one has that the random variables as given by eqs. (C-7)–(C-8) converge at $T\to\infty$ to the left-hand side of eqs. (C-4)–(C-6) and producing thus an ergodic theorem on the equality of ensemble-probability average of the wide sense stationary process $\{X(t), -\infty < t < \infty\}$ and any of its single-sample $\{\overline{X}(t), -\infty < t < +\infty\}$ time average

$$\lim_{T\to\infty} \frac{1}{2T}\int_{-T}^{T} dt f(\overline{X}(t)) = E\{P_{\mathrm{Ker}(H)}(f(X_0))\} = E\{f(X(t))\} \quad \text{(C-10)}$$

$$\begin{aligned}\lim_{T\to\infty} \frac{1}{2T}\int_{-T}^{T} dt\overline{X}(t)\overline{X}(t+\zeta) &= E\{X(0)X(\zeta)\} \\ &= E\{X(t)X(t+\zeta)\} \\ &= R_{XX}(\zeta) \quad \text{(C-11)}\end{aligned}$$

9.9 References

[1] W. Humzther, Comm. Math. Phys. 8, 282 -299, (1968).

[2] Luiz C.L. Botelho, Lecture Notes in Applied Differential Equations of Mathematical Physics, World Scientific, Singapore, 2008.

[3] Y a. G. Sinai, Topics in Ergodic Theory, Princeton University Press (1994).

[4] Luiz C.L. Botelho, Advances in Mathematical Physics, Volume 2011, Article ID257916.

[5] Luiz C.L. Botelho, RandomOper. Stoch. Eq., 301–325, (2010).

[6] O.A. Ladyzhenskaya, The Boundary Value Problems of Mathematical Physics, Applied Math Sciences, 49, Springer Verlag, (1985).

[7] Y. Yamasaki, Measures on Infinite-Dimensional Spaces, Series in Pure Mathematics, World Scientific, vol (5), 1985.

[8] Luiz C.L. Botelho, Random Oper. Stoch Eq. DOI10.1515/Rose 2011-020.

[9] V. Rivasseau, From perturbative to constructive renormalization, Princeton University, (1991).

[10] James Glimm and A. Jaffe, Quantum Physics, a functional integral point of view, Springer Verlag (1981).

[11] Luiz C.L. Botelho Mod. Phys. Lett. B35, 391, (1991).

[12] G. Mackey, The Mathematical Foundations of Quantum mechanics, Benjamin, N.Y., (1963).

[13] A Gleason, J. Math. Mech. 6, 885–894, (1957).

Chapter 10

A Note on Feynman-Kac Path Integral Representations for Scalar Wave Motions

We present Feynman–Kac path integrals representations for scalar wave motions on variable medium. The main new points on them is about their rigorous mathematical validity on the space of continuous functions vanishing at infinity, besides of possesing intrinsically the physical property of causal wave field propagation, thus solving mathematically the long standing problem on the subject of automatically leading to causality wave propagation.

10.1 Introduction

It is well-known that the rigorous Feynman–Kac representaion provides the solution on $L^2(R^N)$ for the diffusion linear problem ([1]). What is not known yet is to generalize the Feynman–Kac path integral representation for hyperbolic problems where the governing partial differential equation is of the second order in time. Althought there have been several attempts to arrive at some mathematically feasible result, none of these has lead to path integrals representations with the causal (retarded) wave propagation already built in on the path integral scheme.

Our aim in this mathematical rigorous methods oriented paper is to explore the Feynman–Kac representation, namely the Trotter–Kato Theorem, but now on the context of the Banach Space $E = C_0(R^N)$, to propose through the use of the Laplace transformation in time; a causal Feynman–Wiener path integral representation for the Cauchy problem associated to the linear wave equation in general medium with a spatially varying index of refraction. These studies are presented on Section 2 and Appendix A. On Appendixes B, C, D, E we present some results which may be of some usefullness for understanding from a calculational point of view, our proposed causal Feynman-Viener path integral for linear wave equation with variables coeficients.

10.2 On the path integral representation

One of the most important results on the subject of path integrals representations for wave propagations are those related to the use of rigorous Wiener measure in infinite dimensional functional spaces ([1]). In this note we intend to present these path integrals results on the context of the Banach functional space of continuous functions on R^N, vanishing at the point ∞: $C_0(R^N) = \{f(x) \in C_b(R^N) \mid \lim_{|x|\to\infty} (f(x) = 0\}$, instead of the usual $L^2(R^N)$ framework of the usual Hilbert Space setting ([2]).

As a first important mathematical remark, we recall that we still have the famous Trotter-Kato product formula in this more ample setting of Banach Spaces as $C_0(R^N)$ for the class of those so called contraction semi groups on Banach Spaces, see ([2]).

Theorem (Trotter–Kato). *Let A and B be the generators of contraction semigroups on a given Banach Space $(E, \|\ \|)$ ([1]). Suppose that the closure on $D(A) \cap D(B)$ associated to the operator sum $A + B$ generates a contraction semi-group on $(E, \|\ \|)$. Then for all $x \in E$ fixed (strong convergence in the Banach Space norm), we have the result*

$$\lim_{n\to\infty} \left\| \left\{ \exp(-t(A+B)) - \left[\exp\left(-\frac{t}{n}A\right) \exp\left(-\frac{t}{n}B\right) \right]^n \right\} (x) \right\|_E = 0 \tag{10.1}$$

An immediate application of the mathematical result Equation (10.1) can be given to problems of diffusion on $C_0(R^N)$. Let us define the "positive Laplacian" Operator $-\frac{1}{2}\Delta$ as the operator acting on the Banach Space $E := (C_0(R^N), \|\ \|_{\text{sup}})$ through the closure of the accretive operator $-\frac{1}{2}\Delta$, originally defined by acting on $S(R^N)$. Now one can show that $\exp(-\frac{1}{2}t\Delta)$ is a contractive semi group on E ([1]). It is another result on the subject that if a given strictly positive bounded continuous function on R^N, $V(x) \in C_b(R^N)$, $(V(x) \geq V_0 > 0)$, one still has that $-\frac{1}{2}\Delta + V$ generates a contractive semigroup on $C_0(R^N)$. As a direct consequence of the Trotter–Kato Theorem applied to the Laplacean operator, one has the Feynman result for the diffudion propagator (for $t \geq 0$)

$$\exp(-t(-\frac{1}{2}\Delta + V))(f)(x)$$

$$\overset{\text{(uniform convergence on } R^N)}{=} \lim_{n\to\infty} \left(\frac{2\pi t}{n}\right)^{-\frac{3}{2}N} \tag{10.2}$$

$$\times \left[\int_{R^{nN}} \exp(-S_n(x;x_1,\ldots,x_n;t))f(x_n)d^N x_n \ldots d^N x_1\right]$$

with the discreticized Euclidean Feynman action

$$S_n(x,x_1,\ldots,x_n;t) = \sum_{i=1}^{n}\left(\frac{t}{n}\right)\left[\frac{1}{2}\left(\frac{|x_i - x_{i-1}|}{t/n}\right)^2 + V(x_i)\right]. \tag{10.3}$$

At this point, one can represent rigorously Equation (10.2) through a Feynman-Kac-Wiener path integral. (By using the Feynman physicist notation for the propagator kernel)

$$\left\langle x\left[\exp\left(-t\left(-\frac{1}{2}\Delta + V\right)\right)\right]y\right\rangle$$

$$\overset{\text{(pointwise)}}{=} \int_{\vec{X}(0)=y}^{\vec{X}(t)=x} \left[D^F(X(\sigma))e^{-1/2\int_0^t (\frac{d\vec{X}}{d\sigma})^2(\sigma)\,d\sigma}\right]$$

$$\times \exp\left(-\int_0^t V(\vec{X}(\sigma))\,d\sigma\right) \tag{10.4}$$

Note that the Wiener measure (with two marked fixed-intercept points (x,y)) is written in the Feynman suggestive form of infinite product of usual weighted Lebesgue measures of sample trajectories connecting the marked points (x,y), i.e. ([1])

$$\overbrace{d^{\text{Wiener}}\mu[\vec{X}(\sigma)]}^{\text{rigorous path measure}} = \left(\overbrace{\prod_{0\leq\sigma\leq t} d\vec{X}(\sigma)}^{\text{symbolic math}} \overbrace{\exp\left(-\frac{1}{2}\int_0^t \left(\frac{d\vec{X}}{d\sigma}\right)^2 (\sigma)\,d\sigma\right)}^{\text{symbolic math}}\right). \tag{10.5}$$

We now intend to apply the above rigorous mathematical results to second order hyperbolic wave motion with datum in $C_0(R^N)$ ([3]). Let us thus write the dynamical scalar wave equation governing our dynamics:

$$\frac{1}{C^2(x)}\frac{\partial^2}{\partial t^2}U(x,t) = \frac{1}{2}\Delta U(x,t) + F(x,t)$$

$$U(x,0) = f(x)$$

$$U_t(x,0) = h(x) \tag{10.6}$$

Here the spatially slowly variable wave field velocity is suposed to be given by the continuous bounded function on R^N; $\frac{1}{C^2(x)} = \frac{1}{C_0^2 q(x)}$, with C_0 denoting a reference wave velocity medium and $1/q(x) := \frac{C_0^2}{C^2(x)} := 1 + \varepsilon\left(\frac{\bar{C}^2(x)}{C_0^2}\right)$ here $\varepsilon << 1$ with $\bar{C}^2(x)$ denoting the "perturbation" wave field velocity $(C^2(x) = C_0^2 + \varepsilon\bar{C}^2(x))$.

Since we are now on the context of Banach Spaces $C_0(R^N)$, one should naturally consider the Laplace transformed problem in relation to the time (and leading thus to time forward wave propagations automatically). It is possible to show rigorously that any causal solution of the Cauchy problem Equation (10.6) $U(0,t)$ is of exponential order in time.

$$\widetilde{U}(x,s) = \int_0^\infty e^{-st} U(x,t)\, dt. \tag{10.7}$$

$$\frac{s^2}{q(x)}\widetilde{U}(x,s) = \frac{C_0^2}{2}\Delta\widetilde{U}(x,s) + \frac{s}{q(x)}f(x) + \frac{h(x)}{q(x)} + C_0^2\widetilde{F}(x,s). \tag{10.8}$$

Let us now note that $\frac{-C_0^2}{2}\Delta + \frac{s^2}{q(x)}$ generates a contractive semigroup on $C_0(R^N)$, with the frequency s being a real number. As a consequence of the above displayed remarks, one has the following operatorial Feynman–Kac path integral representation

$$\widetilde{G}(x,y;s) := \left(\frac{s^2}{q(x)} - \frac{C_0^2}{2}\Delta\right)^{-1}(x,y)$$

$$\overset{\text{pointwise}}{=} \int_0^\infty d\zeta \left\langle x \left| \exp\left(-\zeta\left[\frac{s^2}{q(x)} - \frac{C_0^2}{2}\Delta\right]\right)\right| y\right\rangle$$

$$\overset{\text{pointwise}}{=} \int_0^\infty d\zeta \left\{ \int_{\vec{X}(0)=y}^{\vec{X}(\zeta)=x} D^F[\vec{X}(\sigma)] \exp\left(-\frac{1}{2C_0^2}\int_0^\zeta \left(\frac{d}{d\sigma}\vec{X}(\sigma)\right)^2 d\sigma\right) \exp\left(-s^2\int_0^\zeta \frac{d\sigma}{q(\vec{X}(\sigma))}\right)\right\} \tag{10.9}$$

The above written Feynman-Kac path integral is the result we are looking for as a first step of our study. Note that the complete expression for the scalar wave field is now given through the inverse Laplace transform at least in the Schwartz Distributional sense ([2])[1] (see also Appendix D).

$$U(x,t) = \mathcal{L}^{-1}_{D'(R^+)}(\widetilde{U}(x,s))$$
$$= \mathcal{L}^{-1}_{D'(R^+)}\left\{\int_{R^N}\left[\tilde{G}(x,y;s)\left(\frac{s}{q}f + \frac{h}{q} + C_0^2\tilde{F}(\cdot,s)\right)(y)d^N y\right]\right\} \tag{10.10}$$

It is worth call attention again that since one has a Cauchy initial value problem, one in principle does not have any problem with imposing causality on the obtained wave field configurations. Opposite to previous studies on the subject ([3], [4] and appendix C). Note also that the "operatorial" representation Equation (10.10) is a good candidate for numerical-computer oriented calculations, after time and space suitable discretizations ([4]). (See appendix B).

For the Mathematical oriented readers, one has the rigorous Wiener path integral representation (associated to the diffusion "positive" Laplacean $-\frac{1}{2}\Delta$) for the solution of the wave Equation (10.6) in the frequency domain (see the ref. [1] for Mathematical notation).

The below written rigorous mathematical formulae will not be used in what follows.

$$\widetilde{U}(x,s) = \int_0^\infty d\zeta\left[\int_{C([0,\zeta],\dot{R}^N)} d^{\text{Wiener}}_{(x,\zeta)}\mu[\vec{\omega}(\sigma)]\exp\left(-s^2\int_0^\zeta\frac{d\sigma}{q(C_0\vec{\omega}(\sigma))}\right)\right]$$
$$\times\left[s\left(\frac{f(C_0\vec{\omega}(\zeta))}{q(C_0\omega(\zeta))}\right) + \frac{h(C_0\vec{\omega}(\zeta))}{q(C_0\omega(\zeta))} + C_0^2\widetilde{F}(C_0\vec{\omega}(\zeta),s)\right] \tag{10.11}$$

note that full mathematical conditions on the initial datum leading to each fixed $x \in R^N$ to a s-function which admits non-distributional inverse Laplace transforms remains on open problem as far as we know. Certainly no problem if one choose datum such that $\widetilde{U}(0,s)$ is an analytic function $s = \sigma + i\zeta$ in the half-plane Real $(s) > \infty$, vanishes at $(s) \to \infty$ at all half-planes Real$(s) > \alpha + \delta(\delta \in R^+)$ uniformly with respect to $\arg(s) = \text{arctg}(\frac{s}{\sigma})$, besides the following integrability condition ([2]).

$$\int_{-\infty}^{+\infty} |U(\cdot,\alpha+\sigma+i\zeta)|d\zeta < \infty$$

[1]The definition of the root square of the "positive" Laplacian in R^N is done by the usual (unitary) Fourier Transformation on $L^2(R^N)$, namely: $\sqrt{-\frac{1}{2}\Delta}(f)(x) := f^{-1}[\sqrt{\frac{1}{2}}|k|\hat{f}(k)]$.

Just for completeness of our note, let us write now a non rigorous Feynman path integral representation for our (already built in) casual wave propagation problem (see Equation (10.9) and refs. [3], [4])

$$\begin{aligned}
\widetilde{G}(x,y,s) &= \left(\frac{s^2}{q(x)} - \frac{C_0^2}{2}\Delta\right)^{-1}(x,y) \\
&= -i\left(\int_0^\infty d\zeta \left\langle x \left| \exp\left[i\zeta\left(\frac{s^2}{q(x)} - \frac{C_0^2}{2}\Delta\right)\right]\right| y\right\rangle\right) \\
&\quad \overbrace{-\frac{\pi}{i}\delta^{\text{spectral}}\left(\frac{s^2}{q(x)} - \frac{C_0^2}{2}\Delta\right)}^{=0} \\
&= -i\left(\int_0^\infty d\zeta \left[\int_{\vec{X}(0)=y}^{\vec{X}(\zeta)=x} D^F[\vec{X}(\sigma)]\, e^{\frac{i}{2C_0^2}\int_0^\zeta (\frac{d\vec{X}}{d\sigma})^2(\sigma)d\sigma}\right.\right. \\
&\quad \times \left.\left. e^{-is^2(\int_0^\zeta \frac{d\sigma}{q(\vec{X}(\sigma))})}\right]\right) \qquad (10.12)
\end{aligned}$$

At this point of our note some comments appears to be relevant for analytical manipulations. The first comment is related to the result already expressed by our Feynman-Wiener and Feynman path integrals representations are solutions for obtaining full optics propagation from the idealized geometrical optics, with its idealized rays paths given by the stationary solutions of the geometrical optics Lagrangeans. Namely (see Equation (10.9) and Equation (10.12) respectivelly and chapter twenty, first reference in [3]).

$$\mathcal{L}_{\text{Wiener}}(\vec{X}(\sigma),\zeta) = \left[\int_0^\zeta \left(\frac{1}{2C_0^2}\left(\frac{d}{d\sigma}\vec{X}(\sigma)\right)^2 + \frac{1}{q(\vec{X}(\sigma))}\right)d\sigma\right] \qquad (10.13)$$

$$\mathcal{L}_{\text{Feynman}}(\vec{X}(\sigma),\zeta) = \left[\int_0^\zeta \left(\frac{1}{2C_0^2}\left(\frac{d}{d\sigma}\vec{X}(\sigma)\right)^2 - \frac{1}{q(\vec{X}(\sigma))}\right)d\sigma\right] \qquad (10.14)$$

The second comment is that our path integrals representations open a good deal in principle to understand, through semi-classical and W.K.B. methods, the entire range of interference and diffraction (caustics) electromagnetic scalar (optic) phenomena ([3], [5]) (see appendix A for the wave propagation acoustic case).

As a last comment, we wishe to point out that alternativelly to the path integral representation Equation (9), one has naturally associated to it, the following initial-value diffusion equation on R^N

$$\begin{cases} \frac{\partial}{\partial\zeta}U_s(x,\zeta) = \left(+\frac{C_0^2}{2}\Delta - \frac{s^2}{q(x)}\right)U_s(x,\zeta) \\ U_s(x,0^+) = \delta^{(N)}(x-y) \end{cases} \qquad (10.15)$$

which in some analytical contexts may be as useful as to solve the associated path integral Equation (10.9).

Let us show the last assertive of these first set of results by writing an asymptotic behaviour for our scalar wave field $U(x,t)$ for large times $t \to \infty$ in a compact domain Ω. Firstly we observe that large times $t \to \infty$ the wave field behavior is equivalent to analyze small frequences $s \to 0^+$ on Equation (10.15) and thus large "proper-time" limit $\zeta \to \infty$ on the —"diffusion problem" Equation (10.15) with Dirichlet conditions on Ω, i.e.,

$$U_s(x,\zeta) \overset{\zeta\to\infty}{\sim} e^{E_0(s)\zeta}\phi_0(x,s)\phi_0^*(x,s) \tag{10.16}$$

where $E_0(s)$ is naturally the loweste eigenvalue of the eigenvalue problem on Ω

$$\left(-\frac{C_0^2}{2}\Delta + \frac{s^2}{q(x)}\right)\phi_k(x,s) = -E_k(s)\phi_k(x,s). \tag{10.17}$$

$$\phi_k(x,s)|_{\partial\Omega} = 0.$$

A formal first-order perturbative calculation lead us to the result

$$E_k^{(1)}(s) = \frac{k^2C_0^2}{2} + s^2\hat{C} \tag{10.18-a}$$

$$\hat{C} = \int_\Omega d^N x\left(\frac{1}{q(x)}\right) \tag{10.18-b}$$

$$\phi_k^{(1)}(x.s) = e^{ikx} + \frac{2s^2}{C_0^2}\left[\overbrace{\int_\Omega d^N k' \frac{\hat{Q}(k'-k)}{k^2-k'^2}e^{+ik'x}}^{:=\bar{\psi}_k^{(1)}(x)}\right] \tag{10.18-c}$$

$$\hat{Q}(k-k') = \int_\Omega d^N x \frac{1}{q(x)} e^{i(k-k')x}. \tag{10.18-d}$$

We have thus the straightforward results (with $C := 1 + \hat{C}$)

$$G(x,y,t) \overset{t\to\infty}{\sim} \left(\mathcal{L}^{-1}_{D'(R^+)}\left[\frac{1}{\hat{C}s^2 + \frac{k^2C_0^2}{2}}\right]\right) e^{ik(x-y)}$$

$$+\mathcal{L}^{-1}_{D'(R^+)}\left[\frac{2(\frac{s}{C_0})^2}{\hat{C}s^2 + \frac{k^2C_0^2}{2}}\right](e^{-iky}\bar{\psi}_k^{(1)}(x) + e^{ikx}(\bar{\psi}_k^{(1)}(y))^*)$$

$$= \left(\frac{1}{C_0}\sqrt{\frac{2}{\hat{C}}}\sin\left(\frac{|k|C_0}{\sqrt{2\hat{C}}}t\right)e^{ik(x-y)}\right)$$

$$+ \left(-\frac{\sqrt{2}}{C_0\sqrt{\hat{C}}}|k|\sin\left(\frac{|k|C_0^2}{2}t\right) + \delta(t)\right)$$

$$\times\left(e^{-iky}\bar{\psi}_k^{(1)}(x) + e^{ikx}(\bar{\psi}_k^{(1)}(y))^*\right). \tag{10.19}$$

Here $\vec{K}$ is the wave vector on the bottom of the spectrum of the well-defined free problem on Ω

$$-\left(\frac{C_0^2}{2}\Delta\right)(e^{i\vec{K}\cdot\vec{x}}) = \frac{|\vec{K}|^2}{2}C_0^2(e^{i\vec{K}\cdot\vec{x}}). \tag{10.20}$$

Extensive applications of this direct problem in scalar wave propagation needs the explicity form of the wave field scatter $q(x) = C^2(x)/C_0^2$ and will be left to geophysics oriented papers to appear elsewhere.

Finally, we would like to address ourselves to the important application of using Feynman-Kac path integrals representations to the classical problem of determining the electromagnetic field strength $E[(x,y);z,t]$ originating from a monochromatic point source and propagating in a medium characterized by a deterministic refractive index $\eta(x,y,z)$ in the half-space $R_+^3 = \{(x,y,z) \mid 0 \le z < \infty,\ (x,y) \in R^2\}$. In the well-known paraxial approximation ([6]), this paraxial pulse is supposed to have explicitly the following structural form

$$E_{(K,\Omega)}((x,y);z,t) = \text{Real}\{A[(x,y);z]\,e^{i(kz-\Omega t)}\} \tag{10.21}$$

where the (paraxial) amplitude $A[(x,y),z]$ satisfies the following two-dimensional Schrodinger initial value equation with the depth coordinate z playing the role of the time variable

$$\left(i\frac{\partial}{\partial z} + \frac{1}{2k}\Delta_{(x,y)} - k(1-\eta(x,y,z)\right)A[(x,y);z] = 0. \tag{10.22}$$

Here, the initial date condition is supposed to be known, namely:

$$A[(x,y);z \to 0^+] = B(x,y). \tag{10.23}$$

Instead of the full set Eqs. (10.22)–(10.23), let us firstly solve them under the depth-independent condition on the refractive index $\eta(x,y,z) := \eta(x,y)$. In order to implement such analysis, let us consider the Feynman propagator like problem associated to Equation (10.22) ($k = \frac{1}{2}$, $\vec{r} = (x,y)$)

$$i\frac{\partial\psi}{\partial z} = -\frac{1}{2}\Delta_{\vec{r}}\psi + V(\vec{r})\psi + i\delta(\vec{r}-\vec{r}')\delta(z-z'). \tag{10.24}$$

We added as a mathematical hypothesis the condition that $\psi(z,\ \cdot\ )$ is an analytical function coming from a Fourier transform of a function with compact support (a sort of "frequency" bounded limited signal - [2])

$$\psi(z,\vec{r}) = \mathcal{F}^{-1}_{\omega\to z}[\tilde{\psi}(\omega,\vec{r})] = \frac{1}{\sqrt{2\pi}}\int_{-\Lambda}^{\Lambda} e^{-i\omega z}\tilde{\psi}(\omega,\vec{r})\,d\omega. \tag{10.25}$$

Here Λ is such that supp $\tilde{\psi}(\omega,\ \cdot\ ) \subset [-\Lambda,\Lambda]$.

The operatorial solution of Equation (10.24) is thus given by the mathematical well-defined expression (note now the functional equality on the Hilbert Space $L^2(R^2)$ sense)

$$\tilde{\psi}(\omega,\vec{r},\vec{r}') \overset{L^2(R^2)}{=} +i\left\{e^{+i\omega z'}\left\langle\vec{r}\left|\left[\left(\frac{1}{2}\Delta-(V-\omega)\right)\right]^{-1}\right|\vec{r}'\right\rangle\right\}$$

$$\overset{L^2(R^2)}{=} +ie^{i\omega z'}\left\{\int_0^\infty d\zeta\left\langle\vec{r}\left|\left(\exp\zeta\left(\frac{1}{2}\Delta-(V-\omega)\right)\right)\right|\vec{r}'\right\rangle\right\}$$

$$\begin{aligned}\overset{L^2(R^2)}{=} +ie^{i\omega z'}&\left\{\int_0^\infty d\zeta\, e^{\omega\zeta}\left[\int_{\vec{r}(0)=\vec{r}'}^{\vec{r}(\zeta)=\vec{r}} D^F[\vec{r}(\sigma)]\right.\right.\\ &\times\exp\left(-\frac{1}{2}\int_0^\zeta\left(\frac{d\vec{r}}{d\sigma}(\sigma)\right)^2 d\sigma\right)\\ &\left.\left.\times\exp\left(-\int_0^\zeta V(\vec{r}(\sigma))\,d\sigma\right)\right]\right\}\end{aligned} \tag{10.26}$$

In the depth dependent notation, we re-write Equation (10.26) in the more invariant form

$$\begin{aligned}&\psi[(z,\vec{r}),(z',\vec{r}')]\\ &= i\frac{1}{\sqrt{2\pi}}\left\{\int_{-\Lambda}^{\Lambda} d\omega\, e^{-i\omega(z-z')}\left[\int_0^\infty d\zeta\, e^{+\omega\zeta}\right.\right.\\ &\times\left(\int_{\vec{r}(0)=\vec{r}'}^{\vec{r}(\zeta)=\vec{r}} D^F[\vec{r}(\sigma)]\exp\left(-\frac{1}{2}\int_0^\zeta(\frac{d\vec{r}}{d\sigma})^2(\sigma)\,d\sigma\right)\right.\\ &\left.\left.\left.\times\exp\left(-\int_0^\zeta V[\vec{r}(\sigma)]\,d\sigma\right)\right)\right]\right\}.\end{aligned} \tag{10.27}$$

Since the cut-off Λ is a finite number, one can apply the Fubbini theorem for exchange the (ω, ζ) integrations and obtaining thus the more invariant mathematically rigorous path integral representation for at least with $V(\vec{r}) \in L^2(R^2) + L^\infty(R^2)$.

Now if one takes formally the limit of cutt-off removing of $\Lambda \to \infty$, one could obtain the complete Feynman-Kac path integral representation below written.

Since

$$\lim_{\Lambda\to\infty}\left[-\frac{1}{\sqrt{2\pi}}\left(\frac{e^{-\Lambda\zeta}e^{-i\Lambda\omega}(1-e^{2\Lambda\zeta}e^{2i\Lambda\omega})}{(\zeta+i\omega)}\right)\right] \overset{D'(\mathbb{C})}{:=} \delta((z-z')+i\zeta). \tag{10.28-a}$$

It yields thus

$$\psi[(z,\vec{r}),(z',\vec{r}')] = +i\Big\{\int_0^\infty d\zeta\,\delta((z-z')+i\zeta)$$

$$\times\Big[\int_{\vec{r}(0)=\vec{r}'}^{\vec{r}(\zeta)=\vec{r}} D^F[\vec{r}(\sigma)]\exp\Big(-\frac{1}{2}\int_0^\zeta (\frac{d\vec{r}}{d\sigma})^2(\sigma)\,d\sigma\Big)$$

$$\times\exp\Big(-\int_0^\zeta V[\vec{r}(\sigma)]\,d\sigma\Big)\Big]\Big\}. \tag{10.28-b}$$

If we now introduce a formal parametrization path pure imaginary σ-time (resulting on the so called Feynman "quantum" paths!) through the mathematical formal relationship

$$\vec{R}(\bar{\sigma}) := \vec{r}(-i\sigma) \tag{10.29}$$

one gets the usual operational symbolic Feynman path integral, where we have re-introduced the K-parameter (see the paraxial Equation (10.22))

$$\psi((z,\vec{r});(z',\vec{r}'))$$

$$= \int_{\vec{r}(iz')=\vec{r}'}^{\vec{r}(iz)=\vec{r}} D^F[\vec{r}(\sigma)]\exp\left(-\frac{1}{2}\int_{iz'}^{iz}\left(\frac{d\vec{r}}{d\sigma}\right)^2(\sigma)\,d\sigma\right)$$

$$\times\exp\left(-\int_{iz'}^{iz} V(\vec{r}(\sigma))\,d\sigma\right)$$

$$= \int_{\vec{R}(z')=\vec{r}'}^{\vec{R}(z)=\vec{r}} D^F[\vec{R}(\sigma)]$$

$$\times \exp\left(+iK \int_{z'}^{z} \frac{1}{2}\left(\frac{d\vec{R}}{d\bar{\sigma}}(\bar{\sigma})\right)^2 d\sigma\right)$$

$$\times \exp\left(-iK \int_{z'}^{z} V(R(\bar{\sigma}))\, d\bar{\sigma}\right). \qquad (10.30)$$

For a depth-dependent full case of the potential $V(\vec{r}, x)$, one should just follow R.P. Feynman by introducing the z-ordered product inside the Feynman-Kac path integral Equation (28)

$$\psi[(z,\vec{r});(z',\vec{r}')] = +i\left[\int_0^{\infty} d\zeta\, \delta((z-z')+i\zeta)\right.$$

$$\times \int_{\vec{r}(0)=\vec{r}'}^{\vec{r}(\zeta)=\vec{r}} D^F[\vec{r}(\sigma)] \exp\left(-\frac{1}{2}\int_0^{\zeta}\left(\frac{d\vec{r}}{d\sigma}\right)^2(\sigma)d\sigma\right)$$

$$\left.\times T_\sigma\left\{\exp\left(-\int_0^{\zeta} V(\vec{r}(\sigma))\, d\sigma\right)\right\}\right]. \qquad (10.31)$$

Or in the usual symbolic Feynman path integral notation of Equation (10.30)

$$\psi((z,\vec{r});(z',\vec{r}'))$$

$$= \left\{\int_{\vec{R}(z')=\vec{r}'}^{\vec{R}(z)=\vec{r}} D^F[\vec{R}(\bar{\sigma})] \exp\left(+iK\int_{z'}^{z}\frac{1}{2}\left(\frac{d\vec{R}(\bar{\sigma})}{d\bar{\sigma}}\right)^2 d\bar{\sigma}\right)\right.$$

$$\left.\times T_{\bar{\sigma}}\left(\exp\left(-iK\int_{z'}^{z} V(R(\bar{\sigma}),\bar{\sigma})\, d\bar{\sigma}\right)\right\}. \qquad (10.32)$$

Here $T_{\bar{\sigma}}(\dots)$ means the $\bar{\sigma}$-ordered product operation.[2]

As a pedagogical comment, let us highlight the rigorous mathematical proof of the cutt-off removing of $\Lambda \to \infty$.

In the rigorous mathematical notation of ref.[1] in terms of Wiener path integrals, we define the positive quadratic form in $L^2(R^2)$

$$\langle f, Q^{\Lambda}_{z-z'} g\rangle_{L^2(R^2)}$$

$$:= \frac{1}{\sqrt{2\pi}}\left\{\int_{-\Lambda}^{\Lambda} d\omega\, e^{-i\omega(z-z')} \times \int d\zeta\, e^{\omega\zeta}\right.$$

[2] At this point the reader should not forget that the purely symbolic Feynman path integral is just a formal string of symbols for the real mathematically meaningful object Equation (10.28-b).

$$\times \Big[\int d^{\text{Wiener}} \mu[\vec{r}(\sigma)] f(\vec{r}(\sigma))$$

$$\exp\Big(- \int_0^{\zeta} V(\vec{r}(\sigma))\, d\sigma \Big) g(\vec{r}(\zeta)) \Big] \Big\}. \tag{10.33}$$

One can show that $Q^{\Lambda}_{z-z'}$ is a closable semi-bounded quadratic form. As a consequence there is a set of self-adjoint operators $\mathcal{H}^{(\Lambda)}$ such that

$$\langle f, Q^{\Lambda}_{z-z'} g\rangle_{L^2(R^2)} = \langle f, \exp(-(z-z')\mathcal{H}^{(\Lambda)}) g\rangle_{L^2(R^2)}.$$

By proceeding further, one can shows that $\mathcal{H}^{(\Lambda)}$ converges to a unique self-adjoint operator $\mathcal{H}^{(\infty)}$ in the strong resolvent sense. Now one can see that the Feynman path integral is the famous E. Nelson path integral representation (see second reference in [1]) for the usual Stone Unitary group generated by $\mathcal{H}^{(\infty)}$

$$\langle f, Q^{\infty}_{(z-z')} g\rangle_{L^2(R^2)} = \text{Equation (30)}. \tag{10.34}$$

As a final comment, somewhat oriented to Monte-Carlo (stochastic evaluations) of our proposed Feynman-Kac path integrals representations above written, it is worth remark that by defining a pre-potential $W(x)$, such that it satisfies the first order eikonal partial differential equation

$$\sum_{i=1}^{N} \left(\frac{\partial \Phi}{\partial x^i} \frac{\partial \Phi}{\partial x^i} \right) = V(x) \tag{10.35}$$

where $V(x)$ is a given potential function on $C^1(R^N)$, then we have that the solution of the system of ordinary differential stochastic Stratonovich equations ([4], [7]) [note the all Feynman-Kac path integrals should be associated to the stochastic Stratonovich calculus through calculus the famous Feynman mild-point rule]

$$d_{\text{Strat}} X^i(t) = \frac{\partial \Phi}{\partial x^i}(X^{\ell}(t))dt + \overbrace{dW^i(t)}^{R^N\text{-Brownian drift}}$$

$$X^i(0) = x^i \tag{10.36}$$

defines a diffusion process with (time-invariant) transition density explicitly given by the following Feynman-Wiener path integral ([4], [5])

$$p((x^i,0),(y^i,t)) = \exp(\Phi(x) - \Phi(y))$$

$$\times \left\{ \int_{X^i(0)=x^i}^{X^i(t)=y^i} D^F[X^i(\sigma)] \exp\left[-\frac{1}{2} \int_0^t \left(\frac{dX^i}{d\sigma} \right)^2 d\sigma - \int_0^t V(X^i(\sigma))\, d\sigma \right] \right\}. \tag{10.37}$$

10.3 Appendix A: The Acoustic Case

In this appendix of complementary nature to the bulk of this note, we intend to write path integral representations for scalar wave propagation for more realistic medium of both variable density and variable velocity and modeling physically acoustic scalar wave propagation ([3]) - second reference.

In this more general realistic acoustic case, the governing dynamical wave equation initial value problem takes the following form

$$\frac{1}{C^2(x)}\frac{\partial^2}{\partial t^2}P(x,t) = \frac{1}{2}\Delta P(x,t) - \vec{\nabla}(\ell n\rho(x))\cdot\vec{\nabla}P(x,t) + \frac{F(x,t)\cdot C^2(x)}{K(x)}. \tag{A-1}$$

Here $\rho(x)$ is the spatially varying medium density, $K(x)$ the associated stress-strain bulk modulus (both modeled by strictily positive continuous functions). The dynamical variable scalar wave field is the non-equilibrium pressure field, instead of the usual medium vector position ([5]).

In what follows we are going to treat the more general wave equation initial-value problem, now in the Hilbert Space $L^2(R^N)$ although the point-wise Banach Space $C_0(R^N)$ can be treated straightforwardly as done in the first part of this note, but now with initial datum on the more restrict functional space $C_0(R^N)$

$$\frac{1}{C^2(x)}\frac{\partial^2 U}{\partial t^2}(x,t) = \left[-\frac{1}{2}(-i\vec{\nabla}_x - \vec{a}(x))^2 - V(x)\right]U(x,t) + F(x,t)$$

$$U(x,0) = f(x) \in L^2(R^N)$$

$$U_t(x,0) = g(x) \in L^2(R^N). \tag{A-2}$$

Note that Equation (A-1) can be written in the "Gauge-invariant" form Equation (A-2) by means of the obvious identification

$$\vec{a}(x) = -i(\vec{\nabla}(\ell n\rho))(x)$$

$$V(x) = \frac{1}{2}||(\vec{\nabla}\ell n\rho)||^2_{R^N}(x) - \frac{1}{2}(\Delta\ell n\rho)(x) \tag{A-3}$$

Here all the external imputs $(\vec{a}, V, \ell, g)$ are functional objects on $L^2(R^N)$ and the external source $F(x,t)$ is such that its Laplace Transform in relation to the time t. $\widetilde{F}(x,s)$ for each fixed s - belongs to $L^2(R^N)$.

We have now the standard result ([1], Chapter V) for the inverse Laplace transformed operator kernel

$$\left\langle x\left|\left[\left(\frac{1}{2}(i\vec{\nabla}+\vec{a})^2+V\right)+\frac{s^2}{C^2(x)}\right]^{-1}\right|y\right\rangle$$

$$\stackrel{L^2(R^N)}{=} \int_0^\infty d\zeta \left\{ \int_{\vec{X}(0)=y}^{\vec{X}(\zeta)=x} D^F[\vec{X}(\sigma)] \exp\left(-\frac{1}{2} \int_0^\zeta \left(\frac{d\vec{X}}{d\sigma} \right)^2 (\sigma) \right) \right.$$

$$\times \exp\left[-i \int_0^\zeta \vec{a}(\vec{X}(\sigma)) \frac{d\vec{X}}{d\sigma}(\sigma) - \frac{i}{2} \int_0^\zeta (\text{div } \vec{a})(\vec{X}(\sigma))\, d\sigma \right.$$

$$\left.\left. - \int_0^\zeta V(\vec{X}(\sigma))\, d\sigma - s^2 \int_0^\zeta \frac{d\sigma}{C^2(\vec{X}(\sigma))} \right] \right\} \tag{A-4}$$

which in terms of the original physical medium parameter density can be written as of as

$$G(x, y, s) \stackrel{L^2(R^N)}{=} \int_0^\infty d\zeta \left[\int_{\vec{X}(0)=y}^{\vec{X}(\zeta)=x} D^F[\vec{X}(\sigma)] \exp\left(-\frac{1}{2} \int_0^\zeta \left(\frac{dX}{d\sigma}(\sigma) \right)^2 d\sigma \right) \right.$$

$$\times \left(\frac{\rho(y)}{\rho(x)} \right) \times \exp\left(-s^2 \int_0^\zeta \frac{d\sigma}{C^2(\vec{X})(\sigma))} \right)$$

$$\times \exp\left(+1 \int_0^\zeta (\Delta \ell n \rho)(\vec{X}(\sigma))\, d\sigma \right)$$

$$\left. \times \exp\left(-\frac{1}{2} \int_0^\zeta ||(\vec{\nabla} \ell n \rho)||^2 (\vec{X}(\sigma))\, d\sigma \right) \right] \tag{A-5}$$

here we have used the Stratonovich formula for implementing a needed by parts stochastic partial integration

$$\int_0^\zeta (\vec{\nabla} \ell n \rho)(\vec{X}(\sigma)) \frac{d\vec{X}}{d\sigma}(\sigma)$$

$$= \ell n \rho(\vec{X}(\zeta) - \ell n \rho(\vec{X}(0)). \tag{A-6}$$

It is worth recall that if one had used the It stochastic calculus rule to define the stochastic integral underlying the Feynman-Kac-Wiener path integral Equation (A-5); it could lead to a wrong wave equation under the presence of the vector potential $\vec{a}$ (see first reference-[3]). As a consequence one is really using the Stratonovich prescription for the path integral when involving the presence of an externa vectorial field $\vec{a}(x)$. And even if this is not the case, in other words: if one has solely a scalar potential in the path integral, it does not matter the use of the Stratonovich-Feynman mild point rule or the backward It prescriptions. They are both mathematically equivalent on the realm of Path Integrals.

Another important point to be remarked is that equality on Equation (A-5) must be understood in a weak-sense in $L^2(R^N)$. Namely, for any pair of vectors $(f,g) \in L^2(R^N)$, one has the weak equality

$$\begin{aligned}&\langle f|G(x,y,s)|g\rangle\\&=\langle f|\,\text{Path integral expression on the righ-hand side of Equation (A-5)}|g\rangle.\end{aligned} \tag{A-7}$$

10.4 Appendix B: A Toy model for stable numerics on wave propagation

Let us start this appendix by considering the following velocity medium variable wave equation in one-dimension R^1 (the generalization to the higher-dimensional spaces is straightforward ([2]))

$$\begin{aligned}\frac{\partial^2 U}{\partial t^2}(x,t) &= \frac{d}{dx}\left(C^2(x)\frac{d}{dx}\right)\\U(x,0) &= f(x) \in L^2(R)\\U_t(x,0) &= g(x) \in L^2(R).\end{aligned} \tag{B-1}$$

By re-writing Equation (B-1) in the first order system

$$\begin{aligned}\frac{\partial}{\partial t}\begin{pmatrix}U\\\pi\end{pmatrix}(x,t) &= i\begin{pmatrix}0 & 1\\-\frac{d}{dx}(C^2(x)\frac{d}{dx}) & 0\end{pmatrix}\begin{pmatrix}U\\\pi\end{pmatrix}(x,t)\\U(x,0) &= f(x)\\U_t(x,0) &= g(x)\end{aligned} \tag{B-2}$$

An (weak-sense) operatorial solution in $C([0,t],L^2(R))$ is given throught the Stone theorem ([2]) for each fixed t

$$\begin{pmatrix}U\\\pi\end{pmatrix}(x,t) \overbrace{=}^{C([0,t],L^2(R))} \left\{\exp\left(it\begin{bmatrix}0 & 1\\-\frac{d}{dx}(C^2(x)\frac{d}{dx}) & 0\end{bmatrix}\right)\right\}\begin{pmatrix}f\\g\end{pmatrix}(x). \tag{B-3}$$

However in most computer modelling, the initial datum (f,y) is a somewhat "band-limited" process. For instance, since $-\frac{d}{dx}(C^2(x)\frac{d}{dx})$ is a self-adjoint operator an $L^2(R)$, it has a spectral representation

$$\langle h, -\frac{d}{dx}(C^2(x)\frac{d}{dx})f\rangle_{L^2(R)} = \int_{-\infty}^{+\infty}\lambda\langle h, dE(\lambda)f\rangle_{L^2(R)}. \tag{B-4}$$

As a consequence one should consider all initial datum belonging to a finite-spectral range i.e., there exists $\Lambda > 0$, such that

$$\begin{aligned} E([-\Lambda, \Lambda])f &= f \\ E([-\Lambda, \Lambda])g &= g. \end{aligned} \tag{B-5}$$

The so called finite energy physical initial datum ([2]). For slowly varying medium velocity, one can take the spectral parameter λ as the wave vector k.

Since on these stales sub-space of fixed finite energy, the generator group in Equation (B-3) is effectivelly a bounded operator. So, for short time propagation one has the rigorous result

$$\begin{pmatrix} U \\ \pi \end{pmatrix}(x, t+\Delta t) - \begin{pmatrix} U \\ \pi \end{pmatrix}(x,t) \overset{t\to 0}{\cong}$$

$$i\left(\pi(x,t)\Delta t \left(\frac{U(x+\Delta x,t)+U(x-\Delta x,t)-2U(x,t)}{(\Delta x)^2} C^2(x)\Delta t 2C'(x)C(x)\right)\right.$$

$$\left. \times \left(\frac{U(x+\Delta x,t)-U(x,t)}{\Delta x}\right)\Delta t\right). \tag{B-6}$$

Here

$$U(x,0) = f(x) \tag{B-7}$$

$$\pi(x,0) = g(x).$$

After introducing a discreticized space-time $R^1 \Leftrightarrow (n\Delta)$; $[0,t] \Leftrightarrow (m\delta)$; $U(n\Delta, m\delta) = U_n^m$; $\pi(n\Delta, m\delta) = \pi_n^m$; $U(n\Delta, 0) = f^{(\Lambda)}(n\Delta)$; $\pi(n\Delta, 0) = g^{(\Lambda)}(n\Delta)$; one gets the difference scheme governing the discreticized dynamics

$$U_n^{m+1} = i\delta\pi_n^m + U_n^m$$

$$\begin{aligned} \pi_n^{m+1} =& i\left[\left(\frac{C^2(n\Delta)\cdot\delta}{\Delta^2}\left(U_{n+1}^m + U_{n-1}^m - 2U_n^m\right)\right)\right. \\ &\left. + \left(\frac{2C'(n\Delta)C(n\Delta)\cdot\delta}{\Delta}\left(U_{n+1}^m - U_n^m\right)\right)\right] + \pi_n^m, \end{aligned} \tag{B-8}$$

which is stable in the Von Newmann stability criterium for grid spacements (Δ, δ) satisfying the "Incertanty relationship"

$$\left(\max_{x\in R}|C^2(x)|\right)\frac{\delta}{\Delta^2} < 1 \tag{B-9}$$

$$\Leftrightarrow \Delta t < \frac{1}{(\Delta x)^2 ||C^2(x)||_{L^\infty(R)}}. \tag{B-10}$$

10.5 Appendix C

In this somewhat "advenced calculus" appendix, we show in details how to salve by means of L. Schwartz distribution theory the classical Cauchy problem in R^3 for the scalar wave equation in a wave field case of constant velocity

$$\frac{1}{C^2}\frac{\partial^2}{\partial t^2}U(x,t) = \Delta U(x,t) + \overbrace{F(x,t)}^{\in S(R^3\times R)}$$

$$U(x,0) = f(x) \in S(R^3)$$

$$U_t(x,0) = g(x) \in S(R^3). \tag{C-1}$$

Since the Fourier Transform is a topological isomorphism on the tempered distributional space $S'(R^3)$, one easily obtains the wave field as the $L^1_{\text{loc}}(R^3 \times R^+)$ kernel of a distribution on $S'(R^3)$. Namely:

$$\mathcal{F}_{x\to k}[U(x,t)] := \hat{U}(k,t) \overset{S'(R^3)}{=} \hat{f}(k)\cos(Ct|k|) + \frac{\hat{g}(k)\sin(Ct|k|)}{C|k|}$$

$$-C^2\left\{\int_0^t \frac{\sin(C|k|(t-t'))}{C|k|}\hat{F}(k,t)\right\}. \tag{C-2}$$

Since in $S'(R^3)$ sense, we have the Fourier Transform formulae

$$\mathcal{F}^{-1}\left[\frac{\sin(Ct|k|)}{C|k|}\right] \overset{S'(R^3)}{=} \frac{1}{2C|x|}\{\delta(Ct-|x|)-\delta(Ct+|x|)\} := h_1(|x|,t) \tag{C-3}$$

$$\mathcal{F}^{-1}[\cos(Ct|x|)] \overset{S'(R^3)}{=} \sqrt{\frac{\pi}{2}}\cdot\frac{1}{C|x|}\frac{d}{dt}\{\delta(|x|-Ct)-\delta(|x|+Ct)\} := h_2(|x|,t), \tag{C-4}$$

one gets the usual formulae, after disregarding by hand the "out-going" parts of Eqs. (C-3)–(C-4)

$$U(x,t) \overset{S'(R^3)}{=} (f(x) * h_2^{\text{ret}}(|x|,t)) + (g(x) * h_1^{\text{ret}}(|x|,t))$$

$$-C^2\int_0^t dt'(F(x,t') * h_1^{\text{ret}}(|x|,t')). \tag{C-5}$$

Note that the equality on the above written Equation (C-5) must be taken on distributional sense $S'(R^3)$ (for each test function $P(x) \in S(R^3)$ and for a fixed time t).

It is worth now point out that if the initial datum $f(x)$ and $g(x)$ are respectivelly considered as elements on the Sobolev Spaces $\mathcal{H}^2(R^3)$ and $\mathcal{H}^1(R)$, one may give a point function Mathematical Meaning for Equation (C-2).

$$\hat{U}(k,t) = (|k|^2 \hat{f}(k)) \left(\frac{\cos(Ct|k|)}{|k|^2} \right) + |k|\hat{g}(k) \left(\frac{\sin(Ct|k|)}{|k|^2} \right) - C^2 \left\{ \int_0^t \frac{\sin(C|k|(t-t'))}{C|k|^2} (|k|\hat{F}(k,t)) \right\}. \tag{C-6}$$

Now one has on $L^2(R^3)$, the Fourier Transform inverse formulae

$$\left(\mathcal{O}(a) \overset{D'(R)}{=} \begin{cases} 1 \text{ if } a \geq 0 \\ 0 \text{ if } a < 0 \end{cases} \right)$$

$$\mathcal{F}^{-1}\left[\frac{\cos(Ct|k|)}{|k|^2} \right] = \frac{1}{2C|k|}\sqrt{\frac{\pi}{2}} \frac{d}{dt} \left(\frac{|x|+Ct}{2} - \left| \frac{x-Ct}{2} \right| \mathcal{O}(x-Ct) \right) := j_2(|x|,t) \tag{C-7}$$

$$\mathcal{F}^{-1}\left[\frac{\sin(Ct|k|)}{|k|} \right] = \sqrt{\frac{1}{2\pi}} \frac{1}{C|x|} (1 - \mathcal{O}(|x| - Ct)) := j_1(|x|,t). \tag{C-8}$$

We get thus, the solution on $L^2(R^3)$ for the wave field

$$U(x,t) \overset{L^2(R^N)}{=} \int j_2(|x-x'|,t) \Delta f(x')\, d^3x' + \int_{-\infty}^{+\infty} j_1(|x-x'|,t)((-\Delta)^{1/2} g)(x')\, d^3x' - C^2 \left\{ \int_0^t j_1(x-x',t')((-\Delta)^{1/2} H)(x',t')\, d^3x'\, dt' \right\}. \tag{C-9}$$

Note that if $g(x) \in \mathcal{H}^2(R^3)$, then one obtains a full $L^2(R^3)$ solution, since now

$$\mathcal{F}^{-1}\left[\frac{\sin(Ct|k|)}{|k|^2} \right] = \sqrt{\frac{2}{\pi}} \frac{1}{C|x|} \left[\frac{|x|+Ct}{2} - \left| \frac{x-Ct}{2} \right| \mathcal{O}(|x| - Ct) \right]. \tag{C-10}$$

For the case of initial datum and outputs on space of continuous functions see the Method of Spherical Means in second ref.[2].

10.6 Appendix D: The Causal Propagator – The Retarded Potential

In this short appendix we shall deduce the causal retarded portential in R^3 from our proposed path integral representation Equation (9).

We have thus, the chain of equations:

$$\begin{aligned}
&\tilde{G}(x,y,s)\\
&= \int_0^\infty \left\{ \int_{\vec{X}(0)=y}^{\vec{X}(\zeta)=x} D^F[X(\sigma)] \exp\left(-\frac{1}{2C_0^2}\int_0^\zeta (\frac{dx}{d\sigma})^2 d\sigma\right)\right.\\
&\quad \left.\times \exp(-S^2\zeta)\right\} d\zeta\\
&= \int_0^\infty \left\{ \overbrace{\left(\frac{1}{2\pi C_0^2\zeta}\right)^{3/2} \exp\left(-\frac{1}{2C_0^2\zeta}(x-y)^2\right) e^{-(s^2\zeta)}}^{:=\tilde{G}_s(x,y,s)} \right\} d\zeta\\
&= \left(\frac{1}{2\pi C_0^2}\right)^{3/2} \int_0^\infty \left\{ \zeta^{-3/2} e^{-(S^2\zeta)} e^{-\frac{1}{\zeta}}\left(\frac{(x-y)^2}{2C_0^2}\right)\right\} d\zeta\\
&= \frac{1}{(2\pi)^{3/2}}\cdot\frac{1}{(C_0)^3}\times\left\{2\left(\frac{(x-y)^2}{2C_0^2S^2}\right)^{-1/4} K_{-1/2}\left(2\sqrt{\frac{(x-y)^2}{2C_0^2}S^2}\right)\right\}\\
&= \frac{1}{(2\pi)^{3/2}}\cdot\frac{1}{C_0^2}\cdot\sqrt{2\pi}\cdot\frac{1}{|x-y|}\exp\left(\frac{-\sqrt{2}|x-y|S}{C_0}\right).
\end{aligned} \tag{D-1}$$

Since one has the validity of the above Inverse Laplace Transform on the sense of $D'(R^+)$

$$\mathcal{L}^{-1}_{s\to t}[e^{-as}] = \delta(t-a). \tag{D-2}$$

One obtain the final retarded potential for our wave equation eq(6) on the main text

$$G(x,y,t) = \mathcal{L}^{-1}_{s\to t}(\tilde{G}(x,y,s)) = \frac{\delta(ct-|x-y|\sqrt{2})}{2\pi C_0|x-y|}. \tag{D-3}$$

On the usual case of Equation (60) written with the pure Laplacian $(+\frac{1}{2}\Delta \to +\Delta)$, one obtains the usual result due originally to Linard-Wiechiert

$$\tilde{G}_s(x,y,\zeta) = \left(\frac{1}{2}\right)^{3/2}\left(\frac{1}{2\pi\zeta C_0^2}\right)^{3/2} \exp\left(-\frac{(x-y)^2}{4C_0^2}\right). \tag{D-4}$$

$$G_{\text{ret}}(x,y,t) = \frac{\delta(C_0t-|x-y|)}{4\pi C_0|x-y|} \tag{D-5}$$

10.7 Appendix E: The Causal Propagator – The Damped Case

Let us take into account in the wave equation initial value problem, the existence of a damping term. Namely

$$\begin{cases} \frac{1}{C^3(x)} U_{tt} = \frac{1}{2}\Delta U - \nu U_t + F \\ U(x,0) = f(x) \in C_0(R^3) \\ U_t(x,0) = g(x) \in C_0(R^3). \end{cases} \tag{E-1}$$

Here $\nu \in R^+$, denotes the damping constant.

By proceeding as in previous studies presented in this note, we arrive at the proper-time representation for the problem Green function

$$\tilde{G}(x,y,s) = \left(\frac{1}{2\pi C_0^2}\right)^{3/2} \left\{ \int_0^\infty \zeta^{-3/2} \times \exp[-\zeta(s^2 + \nu s C_0^2)] \right.$$
$$\left. \times \exp\left(-\frac{1}{\zeta}\left(\frac{(|x-y|^2)}{2C_0^2}\right)\right)\right\} = \frac{1}{(2\pi)^{3/2}} \cdot \frac{1}{C_0^3}$$
$$\times \left\{ 2\left(\frac{|x-y|^2}{2C_0^2(x^2+\nu s C_0^2)}\right)^{-1/4} \times K_{-1/2}\left[\frac{\sqrt{2}}{C_0}|x-y|(s^2+\nu C_0^2 s)^{1/2}\right]\right\} \tag{E-2}$$

By noting the L. Schwartz distributional Laplace Transform result

$$\mathcal{L}^{-1}\left[e^{-a\sqrt{s^2+sb}}\right] \overset{D'(R^+)}{=} e^{-\frac{b}{2}t}\left(-\frac{d}{da}\left(I_0\left(\frac{b}{2}\sqrt{t^2-a^2}\right)\mathcal{O}(t-a)\right)\right) \tag{E-3}$$

One obtains the final expression for the causal Green function for the damped case of constant medium wave propagation

$$\tilde{G}(x,y,t) = \frac{A}{|x-y|} e^{-\frac{\nu C_0^2 t}{2}} \left\{ -\frac{d}{da}\left[I_0\left(\nu C_0^2\sqrt{t^2-a^2}\right)\right]\mathcal{O}(t-a)\right\} \tag{E-4}$$

Here the constant A is given explicitly by

$$A = \left(\frac{1}{(2\pi)^{3/2}}\right)\left(\sqrt{\frac{\pi}{2}}\right)\left(\frac{2^{-\frac{1}{4}}}{C_0^2}\right) \tag{E-5}$$

10.8 References

[1] Barry Simon, *Functional Integration and Quantum Physics*, Academic Press, (1979).
– M. Reed & B. Simon, *Methods of Modern Mathematical Physics*, Vol II, Academic Press, (1980).
– Ya. A. Butko, M. Grothous and O. G. Smolyanov, *Lagrangian Feynman formulae for second order parabolic equations in bounded and unbounded domains*; Inf. Dim. Anal. Quant. Probability, Rel. Top V. 13, No 3, 377–392, (2010).

[2] Luiz C.L. Botelho, *Random Oper. Stoch. Equ 18*, 301–325, (2010).
– *Lecture Notes in Applied Differential Equations of Mathematical Physics*, World Scientific, (2008).

[3] L.S. Schulman, *Techniques and Applications of Path Integration*, A. Wiley, Interscience, (1981).
– R.B. Schlottmann, Geophys. J. Int., 353–363, (1999).

[4] Luiz C.L. Botelho, Int. J. Theor. Phys. 49, 1396–1404 (2010).
– *Methods of Bosonic Path Integrals Representations, Random Systems in Classical Physics*, (2006).

[5] Luiz C.L. Botelho, Physical Review 49E (2), R1003-R1004, (1994).
– Mod. Phys. Lett B13 (11), 363–370, (1999).
– J. Phys. A, Math Yen 34 (12), L131–L137, (2001).
– Mod. Phys. Lett 16B (21), 793–806, (2002).

[6] Luiz C.L. Botelho, Modern Physics Letters 14B, No 3, 73–78, (2001).

[7] Eugene Wony and Bruce Hajek, *Stochastic Processes in Engineering Systems*, Springer-Verlag, (1985).
– Zeev Schuss, *Theory and Applications of Stochastic Differential Equations*, Wiley Series in Probability and Mathematical Statistics, John Wiley & Sons, New York, USA, (1980).

Chapter 11

A Note on the extrinsic phase Space path Integral Method for quantization on Riemannian Manifold Particle Motions - An application of Nash Embedding Theorem

Through the use of Nash embedding for Riemann smooth manifolds, we propose a constrained phase space path integral for quantization of one particle motion in Riemannian manifold.

11.1 Introduction

The subject of writing path integral representations for describing the quantum propagation of particles (spinorial or scalar) in Riemannian Manifolds remains as a central theme on the search of consistent framework for the quantization of the gravitation. In last decades, we have seen the appearance of several important studies on the subject. However all those works make full recourses for the intrinsic geometrical properties of the Riemann Manifold where the propagation is supposed to take place (see [1]–[4] for a short sample list of references).

In this note we propose a somewhat different path integral quantization geometrical framework, based on a deep theorem due to Nash that asserts roughly that every Riemannian metric in a given d-dimensional C^∞-manifold $\{M, g_{\mu v}(x)\}$ can be always obtained from an immersion $f^A : M \to R^{S(d)}$ ($f^A \in C^1(M)$ and rank $D_x\ f^A = d$) in a suitable Euclidean space is strictly greater than d ($S(d) \geq 2d - 1$) ([5],[6]). As a consequence one can forseen to write the metric field $g_{\mu v}(x)$ as

$$g_{\mu v}(x) = \sum_{A=1}^{S(d)} \frac{\partial f^A}{\partial x^\mu} \frac{\partial f^A}{\partial x^\nu} \stackrel{\text{def}}{\equiv} \frac{\partial f^A}{\partial x^\mu} \frac{\partial f_A}{\partial x^\nu} \tag{11.1}$$

here $\{x\} \in Dom(f^A)$, a point of $R^{S(d)}$ (containing by its turn the manifold chart of M which x belongs for). Note that eq. (11.1) is a non-linear first order set of Partial Differential Equations for $f^A(x^\gamma)_{\gamma=1,\dots,d}$ with the source term $g_{\mu v}(x)$ given explicitly as input.

11.2 The Phase Space Path Integral Representation

So, let us start our analysis by re-writing the Lagrangian for free-motion on the manifold in terms of the new extrinsic coordinates.

$$\mathcal{L} = \overbrace{\frac{1}{2} M g_{ij}(X^\mu(\sigma)) \left(\frac{\partial X^i}{\partial \sigma} \frac{\partial X^j}{\partial \sigma} \right)(\sigma)}^{\mathcal{L}(X^i, \dot{X}^i)} = \overbrace{\frac{1}{2} M \delta_{AB} \left(\frac{dQ^A}{d\sigma} \frac{dQ^B}{d\sigma} \right)(\sigma)}^{L(Q^A, \dot{Q}^A)} \quad (11.2)$$

where the new particle "extrinsic" coordinates are given explicitly by

$$Q^A(\sigma) = f^A(X^1(\sigma), \dots, X^d(\sigma)) \quad (11.3)$$

with $A = 1, \dots, S(d)$ and we suppose besides that the inverse functional relations hold true (the usual inverse theorem ([2]) of advanced calculus is supposed to be at least locally being under application!). Namely:

$$X^i(\sigma) = G^i(Q^{(A)}(\sigma)) \quad (11.4)$$

plus $(S(d) - d)$ smooth constraints on the motion, when viewed on the non-curved (absolute inertial) referential sistem $R^{S(d)}$

$$\Phi^\ell(Q^B) = 0 \qquad \ell = 1, \dots, S(d) - d. \quad (11.5)$$

At the classical level, the intrinsic free motion in the Riemannian Manifold $(M, g_{\mu\nu}(x))$ as given by eq. (11.2) is entirely equivalent to the classical motion in the extrinsic space $R^{S(d)}$ but now in the full presence of constraints given by eq. (11.5).

We have thus that the Free-Manifold Motion equation

$$\begin{cases} \dfrac{d}{dt}\left\{ \dfrac{\partial \mathcal{L}}{\partial(\dot{X}^i)} \right\} = \dfrac{\partial \mathcal{L}}{\partial X^i} \\ \\ X^i(\sigma) = A^i \quad = \vec{A} \\ \dot{X}^i(\sigma) = B^i \quad = \vec{B} \end{cases} \quad (11.6)$$

is mathematically equivalent to the motion now under constraints ([4]; pages 45-51) in the extrinsic Euclidean Space $R^{S(d)}$. We have thus the new set of Euler-Lagrange equations on the ambient space $R^{S(d)}$.

$$\left(\frac{d}{dt}\left\{ \frac{\partial L}{\partial \dot{Q}^A} \right\} - \frac{\partial L}{\partial Q^A} \right) = \left(\sum_{\ell=1}^{S(d)-d} \lambda_\ell(Q^B, \sigma) \left\{ \frac{\partial \Phi^\ell}{\partial Q^A}(Q^B, \sigma) \right\} \right) \quad (11.7)$$

where the Langrange undetermined multipliers $\lambda(Q^B, \sigma)$ are functions of general co-ordinates $\{Q^B\}$ and of the time σ also. They are determined from the constraints, at least locally in suitable manifold charts

$$\sum_{\ell=1}^{S(d)-d} \left(\frac{\partial \Phi^\ell}{\partial Q^A} \cdot \delta Q^A \right) \equiv 0. \tag{11.8}$$

Note that the set of equations eq. (11.7)–eq. (11.8) must be added with the initial conditions

$$\begin{aligned} Q^A(0) &= f^A(\vec{A}) \\ \dot{Q}^A(0) &= \langle \nabla f(\vec{A}), \vec{B} \rangle_{R^{S(d)-d}} \end{aligned} . \tag{11.9}$$

Let us pass now to our proposed phase-space path integral for quantization of this constraint classical system in $R^{S(d)}$ (eq. (11.2)–eq. (11.5)). So, let us briefly display those basic results on Hamiltonian constraints path integral framework ([5]).

Firstly we consider the classical action functional for such a system. It is given by

$$S = \int_0^T d\sigma \left(\sum_{i=1}^{S(d)} P_A \dot{Q}^A - H(P_A, Q^A) - \sum_{\ell=1}^{S(d)-\ell} \lambda_\ell(Q^B) \Phi^\ell(Q^B) \right). \tag{11.10}$$

The set of variables $\{P_A, Q^A\}$ form the phase space $R^{2S(d)}$ and Φ^ℓ are constraints. They clearly satisfy the Poisson-algebra closeness property (with all Poison-algebra structure constants naturally vanishing in our model). Namely through an explicitly check:

$$\begin{aligned} \{H, \Phi^A\}_{PB} &= \sum_{A=1}^{S(\ell)} \Bigg[\overbrace{\left(\frac{\partial}{\partial P^A} \left(\frac{1}{2} P^B P_B \right) \right)}^{=P^A} \overbrace{\frac{\partial}{\partial Q^A} \left(\Phi^\ell(Q^A) \right)}^{\equiv \nabla \Phi^\ell} \\ &\quad - \left(\frac{\partial}{\partial Q^A} \left(\frac{1}{2} P^B P_B \right) \right) \overbrace{\frac{\partial}{\partial P^A} \left(\Phi^\ell(Q^A) \right)}^{=0} \Bigg] \\ &= \Bigg[\overbrace{\left(\frac{dQ^A}{d\sigma} \right)}^{=P^A_{(\sigma)}} \cdot \nabla \Phi^\ell(Q^A) \Bigg] = \frac{d}{d\sigma} \left\{ \Phi^\ell(Q^A(\sigma)) \right\} = 0 \end{aligned} \tag{11.11}$$

with also the result

$$\{\Phi^A, \Phi^B\}_{PB} = 0. \tag{11.12}$$

Here $\{,\}_{PB}$ denotes the usual operation of taking Poison-Brackets.

An important result on theses quantization frameworks for constraint classical dynamics is that one must fix a sort of "gauge", just in order to assure us that when one is evaluting physical variables $\Omega(Q^A, P^A)$ of the system phase-space along the physical trajectories, they do not depend of the choice of Lagrange multipliers, namely: One must have that

$$\{\Omega, \Phi^\ell\}_{PB} = \sum_\ell d_P^\ell \Phi^p \tag{11.13}$$

together with

$$\frac{d\Omega}{d\sigma} = \{H, \Omega\}_{PB} + \sum_\ell \lambda_\ell \{\Phi^\ell, \Omega\}. \tag{11.14}$$

This gauge fixing procedure can be achieved-according ref.[5] - by introducing a set of surfaces $\chi^\ell(Q^A, P^A) = 0$ on the system's phase space satisfying the conditions below written

$$\text{a)} \qquad \{\chi^\ell, \chi^r\}_{PB} = 0 \tag{11.15}$$

$$\text{b)} \qquad det\left\{\left[\{\chi^\ell, \Phi^r\}\right]_{PB}\right\}_{\substack{\ell=1,\dots,S(d)-d \\ r=1,\dots,S(d)-d}} \neq 0. \tag{11.16}$$

If this is the case, which always happens in our case as can be easily seen, one can perform a canonical transformation which turns the gauge fixing functions $\chi^\ell(Q^A, P^A) \equiv \pi^\ell$ into new canonical moment. Let thus, Q^ℓ be the coordinates conjugate to π^ℓ and Q^*, P^* the remaining set of canonical variables. Note that we can always solve the system $\Phi^\ell = 0$, and find $Q^a = Q^a(Q^*, P^*)$. So the constraint conditions $\Phi^\ell = 0$ and the supplementary conditions χ^ℓ define the new physical phase space Γ^*. And within Γ^*, we have that

$$\begin{aligned} &\pi^\ell \equiv 0 \qquad\qquad \ell = 1, \dots, S(d) - d \\ &Q^a = Q^a(Q^*, \pi^*). \end{aligned} \tag{11.17}$$

The main result ([8]) to be fully used in our study is the following: The matrix element of the quantum mechanical evalution operator is explicitly given by the following phase-space path-integral (we have re-introduced a potential $V(X^i) \equiv \tilde{V}(Q^A)$ into the motion Lagrangian without further mathematical complication)

$$
\begin{aligned}
&\left\langle \left(Q^A_{out}, T\right) \middle| \exp\left(\frac{iH}{\hbar}T\right) \middle| \left(Q^A_{in}, 0\right) \right\rangle \\
&= \int \exp\left\{ \frac{i}{\hbar} \int_0^T d\sigma \left(\sum_{A=1}^{s(d)} P_A \dot{Q}^A - \overbrace{H\left(P_A, Q^A\right)}^{\frac{1}{2}P^B P_B + \tilde{V}(Q^A)} \right) \right\} (2\pi)^{S(d)-d} \\
&\times \left[\left(\det \left[\{\chi^\ell, \Phi^r\} \right]_{\substack{\ell=1,\dots,S(d)-d \\ r=1,\dots,S(d)-d}} \right) \times \prod_{\ell=1}^{S(d)-d} \delta\left(\chi^\ell(Q^A, P^A)\right) \right. \\
&\times \delta\left(\Phi^\ell(Q^A)\right) \left(\prod_{\ell=1}^{S(d)} D^F[Q^i(\sigma)] D^F[P^i(\sigma)] \right) \\
&\left. \times\, \delta^{(S(d))}\left(Q^A(0) - Q^A_{in}\right) \delta^{(S(d))}\left(Q^A(T) - Q^A_{out}\right) \right].
\end{aligned}
\tag{11.18}
$$

Note that it appears (at least in our proposal) that there is not an invariant expression in terms of path integral for the matrix elements asked directly into the original manifold variables $\langle X^i_{out}, T \mid X^j_{in}, 0\rangle$.

In this approach of ours, through the full use of the Nash Theorem, one can do quantum mechanics in Riemann manifolds only when the intrinsic geometrical setting is viewed as usual holonomic mechanical constraints in the extrinsic space $R^{S(d)}$ of absolute embedding frame of the metric manifold.

As useful remark, one can prove that the resulting phase-space path integral eq. (11.19) does not depend on the concrete choice of supplementary conditions $\chi^\ell(Q^A, P^A)$ ([8]).

As useful point in the original Feynman Lagrangean framework, the Feymman propagation is easily re-written in the extrinsic space in the geometrical co-cordinates transformations invariant form as

$$G\left[(Q^A_{out},T);(Q^A_{in},0)\right]$$

$$= \int_{Q^A(0)=Q^A_{in}}^{Q^A(T)=Q^A_{out}} \left[\prod_{0\le\sigma\le T} dQ^A(\sigma)\right](W[Q^\ell(\sigma)])$$

$$\times \left(\prod_{\ell=1}^{S(d)-d} \delta\left(\Phi^\ell(Q^B(\sigma))\right)\right) \tag{11.19}$$

$$\times \left(\exp\left\{\frac{i}{\hbar}\int_0^T d\sigma\left[\frac{1}{2}M(\dot{Q}^A(\sigma)^2 - \tilde{V}(Q^A(\sigma))\right]\right\}\right)$$

with the path measure weight (see eq. (11.4))

$$W[Q^i(\sigma)] = \left\{\prod_{0\le\sigma\le T}\left[\left(g_{\mu\nu}(G^i(Q^A(\sigma))\right)^{\frac{1}{4}}\right]\right\}. \tag{11.20}$$

A simple analysis of the Lagrangean path integral eq. (11.20)–eq. (11.21) shows that analytical disentanglements of them with sensible results on the intrinsic manifold motion appears to be possible only in a Post-Newtonian pertubartive framework of small metrical deviations from the usual Euclidean space: $g_{\mu v}(x) = \delta_{\mu v} + \frac{1}{C}h^{(1)}_{\mu v}(x) + \frac{1}{C^2}h^{(2)}_{\mu v}(x) + \dots$. Here C denotes the light velocity ([6]).

Let us now apply these phase space path integral manifold motion result for the important classical problem of diffusion on the Riemann manifolds, one important problem in Diffusion Physics on Manifolds.

We thus consider the usual linear diffusion equation in the space R^n, but now endowed with a metric $\{g_{ab}(x)\}_{\substack{a=1,\dots,n\\ b=1,\dots,n}}$. Namely ([1]).

$$\begin{cases} U(t,x) = -\frac{1}{2}(\Delta_{g(x)}U)(x) - (VU)(x) \\ U(x,0) = f(x) \in L^2(R^n, gd^nx). \end{cases} \tag{11.21}$$

Here $V(x)$ denotes a real valued function on $\{R^n, g_{ab}(x)\}$, the so called diffusion potential. By following the previous analitically complex time continuation arguments on the connection of the diffusion equation and the Schrdinger equation on R^n ([9]), one can easily writen a path integral (Feynman-Kac-Wiener) for the solution eq. (11.22).

$$U(t,x) = \int_{-\infty}^{+\infty} d^n y\sqrt{g(y}\, K(x,y,t)f(y) \tag{11.22}$$

with the evolution Kernel:

$$K(x,y,t) = \overline{G}(Q_x, Q_y, t)\Big|_{\substack{Q^{x,A}=f^A(x^1,\dots,x^n)\\ Q^{y,B}=f^B(y^1,\dots,y^n)}} \tag{11.23}$$

with $\overline{G}(Q_x, Q_y, t)$ the analitically imaginary time continued path integral equation as given by eq. (11.20), namely

$$K(x,y,t) = G(f^A(x), it); (f^A(y), 0). \tag{11.24}$$

We finally pass to the second quantization problem in a perturbative framework for neutral scalar fields. Here one must given a meaning for the covariant manifold free field path integral under the presence of an external (covariant coupled) field source ([9])

$$\begin{aligned} Z[J(x)] = \tfrac{1}{Z(0)} \times & \left\{ \int \left(\prod_{x^\mu \in M} \sqrt[4]{g(x^\alpha)}\, d\varphi(x^\alpha) \right) \right. \\ & \times \exp\left\{ +\frac{i}{\hbar} \int_M \left[(\varphi(-\frac{1}{2}\Delta_g)\varphi)(x^\alpha)\sqrt{g(x^\alpha)} d^D x \right] \right\} \\ & \times \exp\left\{ +\frac{i}{\hbar} \int_M (J(x^\alpha)\varphi(x^\alpha))\sqrt{g(x^\alpha)} d^D x \right\}. \end{aligned} \tag{11.25}$$

After evaluating the exactly soluble Gaussian Feynman covariant field path integrals and expressing the resulting function determinant by means of the proper-time method, one obtains the "Loop Space" path integral representation for the non-normalized generating Functional associated to the Feynman field path integral eq. (11.26). Namely:

$$\begin{aligned} Z[J(x)] = \det{}^{-1/2}(-\Delta_g) \exp & \left\{ \tfrac{i}{\hbar} \textstyle\int_M \sqrt{g(x^\alpha)} d^\nu x \int_M \sqrt{g(y^\alpha)} d^\nu y \right. \\ & \left. J(x^\alpha)(-\Delta_g)^{-1}(x^\alpha, y^\alpha) J(y^\alpha) \right\} \end{aligned} \tag{11.26}$$

where one has the proper-time motion manifold path integrals representations, see eqs. (11.23)–(11.24).

$$\begin{aligned} \ell g\left[\overset{-\frac{1}{2}}{\det}\left(-\Delta_g \right) \right] &= \int_0^\infty \frac{dt}{2t} \Big(\overbrace{Tr_{L^2(M,g(x)}\left[e^{-t(-\Delta_g)} \right]}^{\equiv K(x^\alpha, x^\alpha, t)} \Big) \\ (-\Delta_g)^{-1}(x^\alpha, y^\alpha) &= \int_0^\infty dt \Big(\overbrace{\langle x, t | e^{-t(-\Delta_g)} | y, o \rangle}^{\equiv K(x^\alpha, y^\alpha, t)} \Big). \end{aligned} \tag{11.27}$$

After re-inserting Eq. (11.26) into Eq. (11.27) we are able to re-write the non-normalized Generating Functional in terms of a dynamics of covariant path integrals ([9]).

11.3 References

[1] N. Ogawa, K. Fuji, A. Kobushkin, Quantum mechanics in Riemann Manifold, Prog. Theor. Phys. 83, 894 (1990).
[2] T. Homma, T. Inamoto, T. Miyazaki, Schrodinger Equation for the Nonrelativistic Particle Constrained on a Hypersurface in a Curved Space, Phys. Rev. D42, 2049 (1990).
[3] N. Ogawa, K. Fuji, The difference of effective hamiltonian in two methods in quantum mechanics on submanifold, Prog. Theor. Phys. 87, 513 (1992).
[4] A Shimizu, T. Inamoto, T. Miyazaki, Path integral quantization of a nonrelativistic particle constrained on a general hypersurface, Nuovo Cim. B107, 973 (1992).
[5] Sternberg, Shlomo, "Lectures on Differential Geometry", Chelsea Publishing, Company, New York, (1983).
[6] Luiz C.L. Botelho, Int J. Theoretical Physics, 48, 1554-1558, (2009).
[7] H. Goldstein, "Classical Mechanics", second edition, Addison-Wesley Publishing Company (1980).
[8] L. D. Faddeev and A. A. Slavnov, "Gauge Fields an introduction to quantum theory", second edition, Addison-Wesley Publishing Company, (1991).
[9] Luiz C. L. Botelho, "Methods of Bosonic and Fermionic Path Integrals Representations: Continuum Random Geometry in Quantum Field Theory, Nova Science, NY, (2009).

Index